Butter Side Up!

Magnus Pyke, after having been born i[...]
educated there at St Paul's School, set s[...]
Canada. He stayed there for seven years and by the time he returned
to do research at London University had acquired a degree in
agricultural chemistry from McGill University. As a boffin in the
Ministry of Food during the war he had something to do with the
nutritional value of the meals served in British restaurants and while
investigating prison diets spent three weeks in Dartmoor.
Throughout twenty-five years as a research director in Scotland he
found time to give lectures, write books and do broadcasts. But it
was only when he came back once more to London to take up his
present post as Secretary of the British Association for the
Advancement of Science that he really came to public notice for his
idiosyncratic exposition of science on TV which won him an award
in 1975. He has a string of degrees and learned qualifications but
wears these distinctions lightly.

Magnus Pyke

Butter Side Up!

or the delights of science

illustrated by ffolkes diagrams by John Stokes

Pan Books London and Sydney

First published 1976 by John Murray (Publishers) Ltd
This edition published 1978 by Pan Books Ltd,
Cavaye Place, London SW10 9PG
© Magnus Pyke 1976
ISBN 0 330 25309 3
Printed and bound in Great Britain by
Hazell Watson & Viney Ltd, Aylesbury, Bucks

This book is sold subject to the condition that it
shall not, by way of trade or otherwise, be lent, re-sold,
hired out or otherwise circulated without the publisher's prior
consent in any form of binding or cover other than that in which
it is published and without a similar condition including this
condition being imposed on the subsequent purchaser

to Bessie
*The most severe yet tolerant
critic of my public performances*

Contents

Sources of illustrations

The title-page drawing, chapter heads and full-page drawings are by Michael ffolkes.

The diagrams on pages 44, 155, 156, 157, 158, 159, 160, 161, 163 and 165 are by John Stokes.

The diagram on p. 169 is by Richard Jones.

Diagrams in text

1 Butter side up!

Ask anyone and they will tell you that if you happen to be having tea and accidentally drop a piece of bread-and-butter, it will always fall on the carpet butter-side down. This assertion, of course, is more a matter of opinion than a matter of fact. In the same vein, a man from foreign parts listening to the conversation in a pub would imagine that it *always* rains on Saturday. For the earnest seeker after truth, this statement is not too difficult to check as the records for rainfall can be obtained from the meteorological office.

Although ready-made statistics may not be available about falling bread-and-butter, it is not too difficult to design an experiment to find out whether bread-and-butter does always fall butter downwards or, indeed, whether there is a real tendency for it to do so. Two conditions are necessary for such experiments to provide meaningful information: they must be properly designed and they must be carried out often enough to make sure that a single result has not happened by chance. For example, a man would mislead himself if, after having tossed a penny twice and having it turn up tails both times, he became convinced that pennies *always* fall tail-side up. He would equally mislead himself if, after holding a piece of bread-and-butter the right way up two inches above the ground, he let it fall: flop. Since under these circumstances it would not have room to turn over, it would inescapably fall right-side up.

In order not to stretch out the excitement too long, it should be said that when an evenly cut and freshly buttered slice of bread is held by its edge about ten feet above the floor and allowed to drop, it falls more often than not butter-side down. If a slice of dry bread were suspended exactly on end and were quite flat on both sides and if it were dropped in still air, it would fall straight down and land on its edge. And when it landed, it would fall over as often one way as the other. This is the sort of thing that happens when a penny is tossed; each time it has an equal chance of landing heads or tails. Since this chance is mathematically fifty-fifty, it may happen that a penny tossed five times falls heads-up five times in a row. But if this happens, the chance of the next five spins giving another five heads is remote (I shall not give the mathematical odds although it is easy to work them out). On the other hand, if a penny does fall heads-up ten times in a row, the chance of heads coming up when it is tossed for the eleventh time is exactly the same as it was the first time it was tossed, namely, fifty-fifty. A good demonstration that the chance of a spun coin falling heads or tails is equal is to toss it a hundred times or, better still, a thousand times. If this is done and the odds are in fact equal, heads (or tails) will gradually show up nearer and nearer to half the time as the number of trials increases.

This brings us back to the flat slices of bread. Under everyday circumstances if a piece of bread is dropped from a good height, say ten or twelve feet, it will tend to turn in the air on to a horizontal plane and then flutter, first one way, then back the other like a leaf floating to the ground. Once again, if the two sides of the slice are marked to identify them, it will be found that it will come to rest on the ground as often one way up as the other. With bread-and-butter, however, things are different.

The layer of butter makes one side of the slice heavier than the other as it is suspended on edge before being allowed to drop. When it does start to fall, therefore, there is a tendency for the butter-side to fall faster than the dry side; this causes the slice to tip and the 'falling leaf' movement to start with the butter-side down. A hundred drops, therefore, do not result in the butter-side being down fifty times and up fifty times. The ratio will be more like sixty times butter-side down and forty times butter-side up. The more often the experiment is done (assuming that it is done properly) the more firmly will the statistics show that there is a true tendency for bread-and-butter to fall in this way. Of course, the exact ratio will depend on how much butter is spread on each slice.

People often say that statistics can prove anything. This is not so; indeed, the position is quite the opposite. Statistics, properly applied, can sort out what is real from what is imaginary when one is looking for something quite small: for example, the push of the revolving earth causing bathwater to swirl, or the direction of fall of a slice of bread-and-butter. Sensible, scientific statistics can save a lot of time and perhaps discomfort as well. Or, on the other hand, they may allow us to take advantage of something we might otherwise have missed.

Two hundred years ago or so it was generally accepted by doctors that the way to treat all manner of diseases, and particularly those caused by the infections from which our ancestors died (but we in this happy age of vaccines and antibiotics do not), was to bleed the patient. Medical museums are full of gruesome relics for 'cupping' and bleeding, and jars and bottles for collecting and weighing the blood drawn. Even more disagreeable to our modern ideas was the use of leeches.

In zoological terms a leech is a predatory or parasite annelid. In common language it can be described as a slug-like creature which lives on blood and attaches itself to its prey (which may be wild animals, cattle, horses, dogs or people) by means of suckers. Before scientific medicine began in about the middle of the last century, many millions of leeches were collected in Europe and America for use in phlebotomy, as medicinal bloodletting was called. Most people consider leeches to be hateful things but many varieties of them are beautifully patterned in diverse shades of browns, reds and greens and have, indeed, been used as designs for the decorations on women's clothes. Some of the coloured leeches can deliver a painful stinging 'bite' when they attach themselves to their prey but for the most part one may not even feel the attack of a leech. The first thing one notices is bleeding from the place after the satiated creature has filled itself with as much blood as it can eat and dropped off satisfied to digest its meal. Perhaps, therefore, it was not so strange – and can actually be thought of as an ingenious example of applied science – for doctors in the past (and for some Oriental physicians to this day) to have thought of the idea of using them.

The idea was remarkable, even gruesome, when it is recalled what terrifying creatures some of the leeches may be. In the Middle East, for example, the young of a type called *Lymnatis nilotica* live in wells and springs and may accidentally get into a drinker's mouth. Worse still, they may scramble into an animal's (or a person's) nose, fasten themselves on to the walls of the air passages and, at best, cause severe bleeding. At worst they can, as they blow themselves up with blood, actually stifle the victim to death: in 1915, Père Faurie, a missionary, was believed to have met his end when blood-sucking land-leeches got into his sinuses, distended themselves with blood and then could not get out again. Oriental land-leeches behave almost as if they were characters in a horror story. They lie in wait on the ground and on the bushes and trees beside the trails through the jungle and then when the first one attacks, the rest gather in swarms to harass and in some circumstances destroy the traveller.

Nevertheless, if blood-letting *is* a useful medicinal treatment capable of making an ill person better, then, rather than having to suffer the pain, minor though it may seem to a courageous patient, of being neatly slit with a scalpel, it might be thought preferable to submit to the sucking of a slug-like annelid. This however, brings us to the main point. While no properly designed statistical trial was ever carried out, evidence gradually accrued: partly it was the ill-effects of blood-letting on already

14

weakened patients; partly it was the absence of effects detectable on patients who were subjected to blood-letting; partly it was observations on patients not subjected to blood-letting but who nevertheless grew better (or died) no sooner or later than anybody else; and partly it was the greatly improved results obtained from new scientific discoveries such as vaccination, inoculation and demonstrably effective drugs – and all this showed that in fact phlebotomy did the patients subjected to it no apparent good.

Scientific truth is very difficult to discover. There are two reasons for this. The first is that nature is, with all its complexity, very obscure. Secondly, there is the number of wrong, nonsensical yet deeply entrenched ideas that exist among currently accepted beliefs. For example, even after the centuries-long belief in the usefulness of drawing blood as a therapeutic measure had been found to be groundless, belief of almost as long duration sprang up in the healing virtues of castor oil. When I was a boy, almost every complaint from fever to gastroenteritis was treated with castor oil. Measles, scarlet fever, mumps and the flu – all were treated indiscriminately with the same specific. Two short generations have passed and today castor oil is hardly ever seen and the health of children and adults alike is not a whit the worse.

But things may sometimes work back to front. Blood-letting and castor oil, accepted for years by people who thought themselves educated as useful agents for the treatment of a wide variety of diseases, turned out when subjected to the rigid discipline of science to be in fact without effect. There have also been instances when what were thought to be old wives' tales have been discovered actually to be founded on truth.

Raspberries have been known for ages, right back to the seventh century and before, and have certainly been cultivated since the early seventeenth century. And for almost equally long old wise women have traditionally recommended raspberry-leaf tea as a specific to ensure easy labour for women in childbirth. But in recent years no one had paid much attention to what the old wives said; that is, no one of any scientific reputation. Then in 1954 three British pharmacists, A. H. Beckett, F. W. Belthle and K. R. Fell, conducted a serious scientific study.[1] They prepared a decoction of raspberry leaves (which is much the same as making raspberry-leaf tea) and carried out careful trials, and when the work was done, the results clearly showed that raspberry leaves *do* contain an active principle that helps women to enjoy easy labour. Trials in the United States and studies by the Oxford Medicinal Plants Scheme gave the same results.

This story is only one argument out of many demonstrating how important it is for the scientific puzzler to use common sense if he is to recognize both nonsense when he sees it and also the truth, no matter how unlikely it may appear to be. And some truths are very unlikely indeed. Take, for example, the report of the man with a grandfather clock which, in its old age, had a tendency to stop. There was no suggestion that it stopped at varying lengths of time after being wound up. It was designed to go for eight days between windings and the man it belonged to always wound it on a Sunday. It had therefore the choice, as it were, of stopping on any day of the week. Since the weight that made it go pulled equally hard throughout the week and the strength of the pull was derived from the force of gravity, common sense seemed to indicate that the day on which it did stop was entirely a matter of chance. The man, however, insisted that more often than not it stopped on a Thursday. And when, after having argued with sceptical friends until he was tired of doing so, he began to keep proper records, lo it turned out that the clock did indeed stop most frequently on a Thursday. The question then arose: how did the clock know it was Thursday?

The answer to the conundrum was only reached after quite a pretty piece of research had been carried out. When this was done, two things were discovered. The first was that since the clock was standing on a carpet, it could be caused to wobble if a force was applied to it, even quite a gentle one hardly amounting to more than a pat. Indeed, even the swinging of the pendulum made the whole clock rock backwards and forwards – not

1 Beckett, A. H., Belthle, F. W., Fell, K. R., *J. Pharm. Pharmacol.*, vi, 1954, 785.

'The question then arose: how did the clock know it was Thursday?'

very much, to be sure, but to a perceptible degree. The next discovery was more interesting. It is well known that the speed with which a pendulum swings backwards and forwards depends on its length. A short pendulum oscillates quicker than a long one. The length of a clock pendulum is of course fixed, which is why it can be used to make the clock keep time. On the other hand, the weight, hanging on its wire, gradually descends during the week, and it may be thought of as a pendulum which, if it were made to swing, would start going backwards and forwards quickly and then gradually swing slower and slower. And in the clock that showed a tendency to stop on Thursday it happened that by the time Thursday came round the length of wire was such that the tiny wobble imparted by the pendulum was exactly in step with the swinging of the weight, which did then indeed start to swing. The weight being heavy caused the wobble to become worse and worse until is was enough to upset the pendulum and the clock stopped.

Science has achieved many marvellous things. Out of it have come transistors and television tubes, vaccination and the Pill, atomic bombs that actually go off and people who *don't* die of tuberculosis any more. But there are certain things it cannot do (I shall discuss some of these later) and there have been occasions when scientists, even when they have thought themselves to have been most scientific, have been wrong. Let us consider Popeye and the can of spinach.

Years ago, people thought little of spinach: 'It is cold and moist,' wrote a chap called Lovell way back in 1659.[1] 'It yieldeth little nourishment . . . and easily causes vomit.' And a writer called Turner[2] put it like this: 'Spinach . . . is seldom used in physick, and I believe not very substantial food, though some greedily eat it.' And more than a hundred years later, in 1799, another expert called Willick[3] was still saying that spinach 'affords no nourishment, . . . weakens the alimentary canal . . . [and] is a proper food for the weak and debilited.'

1 Lovell, R., *Panboranologia, or a compleat herball*, Oxford, 1659.
2 Turner, R., *Botanologia, the British physician; or the nature and virtue of English plants*, London, 1687.
3 Willick, A. F., *Lectures on diet and regimen*, London, 1799.

In due course the scientists came along. First of all they analysed spinach and found that it contained calcium and iron. Of course, all sorts of other foods also contain calcium and iron and in greater amounts than spinach does. Nor did the analysts take the trouble to find out whether the calcium and iron spinach does contain is any good to the people who eat it. In fact, it is highly doubtful whether it is useful at all because spinach also contains quite a lot of another substance, oxalic acid; this combines with calcium and iron to prevent the body getting hold of them even after the spinach has been eaten.

People often have mistaken ideas like this today. They often seem to hanker after 'protein' foods and go and buy soya beans and peanut butter to top up their protein intake. Peanut butter and soya beans do contain protein, to be sure, but then so does bread and almost every other staple food one eats, except sugar and fat. Furthermore, the sort of people who can afford to bother with soya beans and peanut butter from a 'health-food' shop are almost certainly getting about twice as much protein in their ordinary food as they need anyway. And in the 1930s the same sort of confusion arose in the public mind. 'What does calcium do?' they asked the experts. The experts were fearful of not being able to explain how calcium is absorbed through the digestive machinery into the blood, where it performs a variety of functions, including being incorporated in the bony framework structures that keep our whole body from collapsing into a heap; they merely said briskly, if perhaps a whit inaccurately, 'Calcium is body-building.' 'If only I eat enough spinach,' all the puny, hollow-chested little men said to their reflections in the mirror, 'it will build my body' (as if by eating enough bones, a poor weak mongrel greyhound could turn itself into a bulldog or a Great Dane).

As if this were not enough, in the 1930s it turned out that spinach contained vitamin A activity and vitamin C; there was no holding the faithful belief of the multitude in the magical properties of this vegetable. No one noticed that all other green vegetables also contain vitamin A activity and vitamin C. Within twenty years the area of land in the United States devoted to the farming of spinach increased from 5,000 to 105,000 acres. In the 1930s a cartoonist, Max Fleischer, invented Popeye the immortal sailorman and his girlfriend Olive Oyl. From that moment it was fixed for all time in the public mind that when a man was exhausted after a day's work or looked like being bested in a fight by a bully twice his size, all he had to do was swallow a can of spinach and strength and vigour would immediately be restored tenfold.

Everyone knows that the cartoons about spinach are just a joke. At the same time it is generally believed that spinach *is* a nourishing food and *does* contain specially large amounts of vitamins and minerals. It is true, as I have said, that spinach contains vitamins A and C and iron and calcium as well. In this respect, however, it is not outstandingly different from spring greens or turnip tops. Strangely enough, it is not only the non-scientists who have given spinach far more credit than it deserves. In 1924 the US Surgeon-General referred to four separate reports in praise of spinach. In 1931 his reports referred nine times to the calcium, iron and vitamins in spinach. Little wonder that spinach, whether fresh, canned, frozen or sieved was esteemed as a specially valuable article of diet. It was even sold in powder and tablet form by chemists on the basis of its pretended therapeutic virtues.

No one would enjoy playing golf or snooker if it was easy. It is because it is so difficult that it is delightful. And the same holds for science. It was a hard job to discover that it was necessary to eat vitamin C to avoid suffering from scurvy. In fact it took at least 300 years to convince the medical and scientific people that this was so. Even then it took more than twenty years again before the chemical nature of vitamin C and how it operates were worked out. Great scientists possess very special skills – as do great golfers – for being able sweetly to hit the target. No matter that even when they have shown us the way, we still have difficulty in doing as well as they do. Consider the spinach business a little further.

It was only a short while ago that Dr R. Hunter [1] raised yet another point, almost exactly the opposite of the Popeye belief in the virtues of spinach. This idea was that spinach is so full of vitamins – and this time he was talking about a comparatively

1 Hunter, R., *Lancet*, i, 1971, 746.

new vitamin called folic acid – it might be bad for children. He pinned his idea on the fact that under some circumstances, if laboratory animals eat too much folic acid, they become tense and irritable. Is it possible, he thought, that when Popeye (or anyone else) swallowed a can of spinach, the extra folic acid, rather than making him strong, made him irritable and pugnacious? What mother, urging her little Johnnie to eat up his spinach, wants him to become tense and quarrelsome?

One further point about spinach: in Japan they do not run their blood-transfusion service as we do on the voluntary contributions of people like ourselves who, for a biscuit and a cup of sweet tea, are prepared to help the good work. For the most part blood in Japan is provided by commercial blood-sellers who for a fee supply what is needed as an article of trade.[1] Some of these people, known locally as *takos* (the Japanese word for octopus), who run their business at a high level of productivity, supply up to two bottles of blood a day. And having heard about Popeye even in far-off Japan, they make a point of eating spinach in the belief that it will 'thicken' their blood. It is true that they eat a lot of dried sardines as well. Just the same, it would be good science to do some statistics on *takos* who eat spinach and on those who do not to see whether its reputed virtues actually exist.

There are a lot of things that science has discovered. Some of this knowledge is directly useful; for example, the chemistry of aspirin and vitamin A is known, as are the ways to make them in the laboratory. Yet, as the spinach story illustrates, at least for vitamin A, this may not answer any further question of whether eating more vitamin A than one is getting anyway will do you any good. To find this out more experiments may be needed. And for lack of such experiments both scientists and non-scientists have made terrible blunders – and will no doubt go on doing so.

In the 1950s some rather peculiar experiments suggested that chlorophyll, the green colouring matter in leaves, when used as a kind of poultice, had some effect in checking the multiplication of certain bacteria which might grow on infected wounds.

1 Anon, *Transfusion*, iii, 1963, 213.

Within the space of a few months the following remarkable chain of reasoning had been strung together: because chlorophyll checked the growth of these bacteria and because bacteria (not necessarily this special sort but all kinds of bacteria) may produce compounds that cause infected wounds to smell, it followed (so ran the argument) that chlorophyll would stop not just infected injuries but people themselves from smelling. At once a rash of chlorophyll tablets (to stop the breath from smelling), chlorophyll toothpaste (to stop the teeth from smelling), even chlorophyll socks (to stop the feet from smelling), appeared on the market.

Science is a remarkable system of thinking. Perhaps this is why it frightens some people while others believe that it is more remarkable than in fact it is. The point to cling to, however, is that remarkable as science may be and dramatic as are many of the things that have come out of it (from aeroplanes that fly faster than sound to plastic washing-up bowls), nevertheless it is only common sense – glorified and systematized common sense, if you like, but derived from human heads just the same. And common sense tells us that we have been eating chlorophyll for years. After all, it is what makes spinach so green. And we know that, regardless of the vitamins and minerals in it and regardless of what it can or cannot do to help Popeye the sailor-man win his battles, it has nothing to do with whether's one's breath smells or not.

The words of the Duke of Wellington come to mind. To a gentleman who accosted him in the street saying, 'Mr Jones, I believe?' the Duke, marching with dignified steps into his club, is alleged to have replied, 'Sir, if you believe that, you will believe anything.' Perhaps, however, an anonymous poet should be allowed to have the last word. He wrote:

The goat that reeks on yonder hill,
Has browsed all day on chlorophyll.

2 In autumn when the shells are brown

There cannot be many people who, after lifting off the top of a boiled egg and looking inside, have not, at one time or another in their life, said, 'This egg looks funny.' Why, you may ask, do they call it funny? They might reply, had they heard the question, 'Because it's green' (or pink, or blue, or black, or mauve). Very well, admitting that people would describe a boiled egg as 'funny' if its colour were different from what they expected it to be, the core of the question remains, why are people put off by the colour of what they eat and particularly by an unexpected colour?

This brings us to that widely distributed although anatomically peculiar individual, the man of whom it is said that his eyes are bigger than his stomach. This means, of course, the individual who, when he sees food that looks unfunny (or perhaps one should say familiarly attractive) wants to eat it. Such a person expects potato chips to be golden brown. If an artistically minded chef served up chips that were bright blue, nobody would eat them. And even if a group of blind men sitting at the corner table made particularly favourable comments on how delicious they were, the sighted diners accustomed to brown chips would still refuse to eat blue ones.

It is hard to say whether there is some inherent beauty and attractiveness in any particular colour applied to food or whether the attraction, where it exists, is related to familiarity. Take the common predilection for brown-shelled eggs. This taste which, while not universal, is quite widespread is particularly odd when considered by the judicious mind. The colour of the shell has little or nothing to do with the flavour of the egg inside it. Shell colour is a hereditary characteristic relating to the breed of the hen. Of itself it has no dietary or culinary virtue whatever. After all, in consuming an egg one does not eat the shell. It is just as easy to produce free-range white eggs and brown eggs from battery hens as it is to produce brown eggs from free-range hens and white eggs from birds kept in batteries. For that matter, the eggs from battery birds, provided they are appropriately fed, may taste more delicious and interesting than those of hens running around a yard. Indeed, on a badly run farm where the hens are allowed to lay all over the place, eggs may be picked up in peculiar places weeks after they are laid and be stale or even downright bad before they are sold.

The liking of British consumers for eggs with brown shells is quite mysterious. Consider the situation in the United States. For many years, the only eggs that could be marketed in Boston in the State of Massachusetts were brown ones while down the line in New York only white eggs would be accepted. Originally there had been a rational explanation for this diversity in taste. Once upon a time chickens kept by the farmers in Massachusetts belonged to those breeds that laid brown eggs whereas farmers in the State of New York kept mainly white-eggshell-laying hens. It followed, therefore, that housewives in Boston felt some confidence that when they bought brown-shelled eggs they had been produced locally and consequently stood a good chance of being fresh. On the other hand, white-shelled eggs had probably been shipped in from New York or even further afield. And the New York housewives believed that brown-shelled eggs had come from a distance and were consequently less likely to be fresh than white ones.

It is curious to find that, while the beliefs of Boston and New York customers two or three generations ago when eggs were collected in buckets and shipped by train in goods wagons probably possessed some validity, the same beliefs – or at least half of them – persist to the present day in places far remote from either New York or Boston and at a time when the term 'fresh' or even 'new laid' possesses hardly any meaning. We know perfectly well that when there is a glut of eggs or a strike of transport workers or any of the hundred and one events with which a complicated industrial community like ours has to deal every day, eggs can quite safely be left in cold storage and be brought out quite 'fresh' anything up to a year later.

Whichever way you like to take it, love of brown eggshells which nobody eats is a mysterious emotion. At first blush, one might however argue that a preference for the yolk of any egg which is to form part of the sacred combination of egg-and-bacon at the British breakfast to be bright yellow or possibly reddish-yellow has some basis in logic. Eggs with strongly coloured yolks, it might be said, have more flavour or perhaps even a better flavour than eggs with pale yolks. And, after all, one does actually eat the yolk of an egg whatever may happen to the shell. This argument carries less weight, however, when one considers some of the facts that are being brought to light by scientists who are trying to measure likes, dislikes and preferences and study what really are the factors that affect the way people choose their food.

It is a common saying that nobody bakes a cake today the way mother used to make it. It does not matter when 'today' is or what kind of a cook mother was – it probably made no difference to the fond memories of men and women long since grown up if mother's cooking was terrible – the memory of tastes long past is almost always sweet. But if we assume that mother was a good cook in those long-forgotten days, her cake was probably a mixture of flour, sugar and eggs (we will say nothing of the fruit and peel and all those other adventitious ingredients). And the more eggs she put into the cake-mix, the yellower was the final product that eventually came out of the oven. Time passed on and mothers no longer felt inclined to spend their time slaving, as the expression has it, over a hot stove baking cakes. Neither were they inclined to spend more money than they needed to spend on buying cakes from the manufacturers who, consequent on the withdrawal of mothers from their domestic labours, were now undertaking the manufacture of cake.

The upshot of all this has been that manufacturers, who cannot afford to take the risks that mother took of their cakes falling in or coming out burnt, studied the scientific principles which govern the way a cake does behave. For example, the quality of the flour which is measured, not only for the precise amount of the protein, gluten, it contains but for the exact composition of the gluten, has an enormous effect on the evenness of a cake, whether the bubbles in it are regularly distributed throughout the slice or whether one or two big bubbles form. The exact amount of fat and the detailed chemistry of whatever fat is used, the amount of sugar and the precise size of its grains, the amount and character of egg – whether it is fresh egg, dried egg or an egg substitute which may not be egg at all – all these are important and have a major influence on whether the manufacturers' cakes are all equally good or whether some are good and some less good or, indeed, whether the whole lot are no good at all. Of course, only if the cakes are consistently wholesome and good will the customers buy and enjoy them and the baker stay in business.

But besides understanding their business in this way, cake manufacturers must also understand their customers. The first thing they must learn is that no matter how excellent are the ingredients they use or how delicious and attractive are the cakes, no one will buy them, and in consequence, no one will benefit from all those rich nourishing ingredients, if the cakes are too expensive. One of the things manufacturers learned to do in their efforts to keep prices down was to make cakes with perfect crumb structure, agreeably sweet and tasty, yet without any eggs, which were often the most expensive ingredient they had to buy. They soon found, however, that although these cakes might be popular with customers who were blind – or at least colour-blind – or with people who ate their tea in dark glasses, they were not acceptable to most ordinary cake-eaters because, there being no egg among the ingredients, when cut they were white inside.

Nevertheless these cakes were inexpensive and in every other respect than colour they were attractive – their taste was good, they were light and fresh and of an admirable consistency. The obvious step was, therefore, to add some colour to the mixture. This has been done in traditional cooking for ages. For example, in Spain rice is commonly coloured with saffron. What objection could therefore be raised in a modern scientific society to adding colouring matter to give the cake the yellow colour that the customers found attractive. Surely this was merely a harmless cosmetic? It could logically be argued that if housewives could add to their attractiveness by using the benefits of applied science to colour their cheeks and lips red and their eyelids blue by means of dyes which made no pretence to be natural to the pigmentation of the skin and consequently fooled nobody, why should not they (or the manufacturers who served them) use similar means to add to the attractiveness of their cakes? The dyes used to colour cakes yellow were most rigorously tested to ensure their harmlessness. They were tasteless and odourless. And although the cakes containing them were as yellow as the old-fashioned cakes had been that were made with egg, nobody *said* that they did contain egg. All that the colouring achieved was to give the cakes the colour that people had got it into their heads cakes ought to have. They would have been just as wholesome and every bit as tasty if they had been coloured purple but most people have the feeling that cakes ought not to be that colour.

When fashion demands that ladies' fingernails should be bright red we soon get used to seeing them that colour. Indeed, we should probably feel that the ladies were somehow in a rather undressed condition if the ends of their fingers were any other colour. We have all had the experience of waiting patiently downstairs while the mistress of the house was upstairs 'putting her face on'. The situation with food cosmetics, however, is slightly peculiar. Chocolate cake, for example, is popularly supposed to be the particular shade of brown we describe as 'chocolate colour'. It is, however, perfectly feasible to produce a wholesome and palatable cake (possibly the very same cake coloured the attractive 'egg-yolk' yellow we have just been discussing) but this time coloured chocolate brown. People taking it from a supermarket shelf will think nothing odd in this nor will those eating it for tea consider that there is anything funny about it. 'Give me another slice of that chocolate cake, Mother,' they will ask because they like the flavour. But the flavour may not be that of chocolate at all. On the contrary, chocolate-coloured cakes have proved to be every bit as popular when they are flavoured with vanilla! Nor does the brown dye that gives the cake its colour come out of a cocoa bean: why should it? That lovely pink glow on the smiling female face opposite does not necessarily come from the generous circulation of red blood below the surface. We think no less of its beauty if it comes from a chemical works and is applied on the surface.

If there is nothing funny about a brown cake the colour of chocolate and tasting of vanilla, would there be anything odd in a brown fat the same colour for smearing on dry bread? If you like, this brown spread could be flavoured to taste of butter.

The butter story is a very peculiar one. To start with, no one has ever produced any satisfactory answer to the question of why it is we like to smear a stiffish, softish, yellow fat called butter on slices of bread. In India, where the weather is hotter than in the so-called Temperate Zone where most industrialized butter-using societies are to be found, the traditional form of butterfat is an oil called ghee. Obviously this presents no spreading problem. So far, however, butter-using communities would consider it 'funny' to have to pour their butter out of a jug. And they do like their butter to be yellow. Thus it happened that when margarine was first invented in 1868 by a Frenchman, M. Hippolyte Mège Mouriès, who discovered a way of making something rather like butter out of lard, one of the first things the dairymen did in their efforts to stop its being a commercial success was to get the government to pass a regulation decreeing that margarine could only be sold provided no colouring matter was added to it. There was no real excuse for this because, while some of the butter one buys in the shops derives its yellowness solely from the cream it was made of, much more owes a large part of the yellow colour we customers have come to expect to colouring matter which has been added to it. Just the same, even though the white margarine was every bit as nourishing and tasty and spread just as well as yellow margarine, people did not fancy it – even when the producers sold it with a pinch of yellow pigment in a twist of paper, rather like the pinch of salt the potato-crisp people used to put into each of their bags. And the same thing happened when one of the States in America insisted that margarine there should only be sold if it was dyed blue. The unkindest cut of all was administered by the farming community in Germany who proposed the decree that margarine should only be sold so long as it was coloured the exact shade of the brown varnish on the wainscot of the walls of the Reichstag council chamber! Brown eggs – yes; but brown margarine – no, or at least, not yet.

Perhaps the time will come when circumstances will force us to change our ideas about what is funny about food. A cow grazing on an acre of field will produce a ton of milk or more in a year. If one were to set up a series of electrodes all over the field and pass a current from one electrode to another so that it travelled through the soil, all the worms in it would come to the surface where they could be harvested. And an average acre would be found to contain a ton of worms. There is nothing wrong with worms as food. Fish eat them every day. They are – even if a trifle gritty – full of protein and readily digestible. Yet the average supermarket customer would almost certainly consider them to be unacceptable. There is no real reason for this. As I have said, their chemical composition implies that they would be nutritionally satisfactory. Furthermore, funnier things have been eaten by various races of our fellow men in diverse parts of the world: insects, for example.

We eat without complaint the sticky regurgitation of an insect – the bee. Indeed, there are people who, without any good reason, attach nutritional excellence to honey, which is hardly warranted in the light of cold scientific fact. Why then would the readers of this book turn up their noses with disgust if they were offered caterpillars or even the next stage in caterpillar development, pupae, as a dietary ingredient? The answer seems to be the simple one that up till now people have not been accustomed to the idea. Yet the first thing a visitor to a Japanese silk factory will notice is the delicious smell of frying pupae. When the girls working in the factory have unwound the fine skein of silk from the silkworm cocoon, they fry up the grub inside. They are allowed to make use of the insects in this way as one of their 'perks' just as the workers in a Scottish distillery are allowed a 'dram' of whisky when they knock off work.

What about ants? Surely, these are clean, tidy, diligent creatures which operate a well-ordered socialist society on good democratic principles? Why, therefore, should we think it 'funny' – or even downright disgusting – to eat them? Yet in Australia the Aborigines consider a particular species of ant to be a delicacy. This is the sugar ant. These creatures store sugary nectar in their abdomens until they swell up into quite a bloated ball. The local inhabitants snatch them up as they run, hold them firmly by the front end and then, while they are still struggling, bite off their abdomens and swallow them down. What is particularly appealing, or so the Aborigines say, is the delicious sweetness as they crush the back end of the ant in their teeth, followed by the sharp stinging flavour of formic acid, which is the characteristic ingredient of ants.

Rabbits, if left to themselves, will consume anything that grows, and we think nothing of eating the rabbit as our food before he has despoiled in foraging for *his* food half the crops that were growing for ourselves. Yet we might hesitate to deal in the same way with locusts. There are, however, markets in a number of African towns where locusts are sold as a delicacy. They can be purchased either raw or fried. And when fried, crisp and brown, they are exceedingly tasty, possessing a flavour midway between shrimps and whitebait – and crunchy to boot.

For a group of people who think nothing of swallowing practically live oysters whole, we could be thought pretty squeamish for the way we turn up our noses at insects – nay more, people have been known to raise violent objection to the presence of even a single beetle in a slice of veal-and-ham pie.

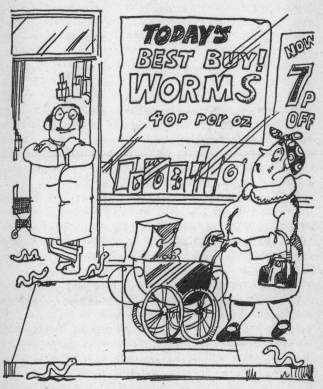

'Yet the average supermarket customer would almost certainly
consider them to be unacceptable.'

North of Edmonton in the Canadian province of Alberta, which is currently booming with prosperity from the exploitation of natural gas and oil, are the Athabasca tar sands which may one day prove to be one of the biggest deposits of fuel in the world. Perhaps, however, we may also benefit from the example of the Dog Rib Indians who come from this part of the world and, for all we know, may still be there when all the oil is gone. These people not only eat the maggots of the warble flies which lay their eggs in the hides of their caribou but consider them a delicacy. The children are particularly fond of them. They pull them out of the caribou's skin and eat them raw alive. They say the taste is as good as gooseberries, and they are much better and more nourishing than the sweets we give our children.

Perhaps if we do not fancy the idea of warble-fly maggots as food, we would take more readily in these hard times to caterpillars. The Pai-Uti Indians, one of the less well-known tribes who live along the California-Nevada border, at one time (before they acquired a taste for fried chicken) had a booming industry for the collection and preparation of caterpillars, which in the 1920s were eaten in large quantities. While over the border in Mexico not only are caterpillars cut out of the leaves of certain local plants and sold as a local speciality but in the same region the famous *ahughutl*, a cake made out of water-bug's eggs, has a ready-made custom at the markets.

How strange it is that, for no logical reason, we consider insects 'funny' when they offer so tasty and various a source of food. In the tropics there is a creature, *Rhynchophorus palmarum*, the palm-worm which grows inside the trees until it becomes as large as one's thumb. The local inhabitants put their ear against the trunk until they hear it moving about, cut down the tree and chop out the worm (it is not, in fact, a worm at all). 'They cook it as we cook cauliflower,' wrote one traveller.[1] 'The taste is more agreeable than artichoke.' And we let all this go by untouched. Yet insects have been royal food: 'Locusts,' wrote Father Comboue in 1886, 'are to be found at the royal table at Tanayariva. The late queen Ranavolona II kept, in addition to her hunters and fishermen, some women who merely

1 Merian, Maria Sibylla, *General history of the insects of Surinam*, 1771.

scoured the fields to collect locusts and other grasshoppers so that the palace should never go short.'

Lots of people keep rabbits and, in due time, eat them one by one or send them to the market in batches to be eaten by other people. It is only since the British as a nation became richer and lived in better houses with wall-to-wall carpeting and a garage for an inanimate car instead of a backyard for livestock that the custom of keeping rabbits and, perhaps, a few chickens as well, has tended to die out. If the nation should become poorer in the future, we may find more people once again going in for this simple process for converting old cabbage leaves into meat – and the occasional pair of fur gloves. Why is it, one may ask, that people are prepared to use rabbit-meat as food when they would recoil in horror if anyone suggested that they should eat that other familiar domestic animal, the dog? The answer does not lie, as many might think, in the fact that rabbits are native to the British Isles and consequently familiar to the British. As a matter of fact rabbits were introduced into Great Britain some-where around the thirteenth century when they were probably brought from Spain. In any event, the argument for familiarity is not a good one since we quickly started to eat and enjoy turkeys, which were entirely novel when they arrived in ships from America; and the dog – one must repeat – which is a familiar beast, has never been a popular item of diet here. What makes this odder still is that Captain Cook, when he was given dog-meat stew when exploring north of Australia pronounced it excellent, tender and with a flavour reminiscent of chicken.

I am not suggesting that we should eat dogs, merely pointing out that when scientists talk about food and nutrition, the protein shortage, slimming and the vitamin content of orange juice or carrots, these facts – important though they may be – must take second place if people refuse to eat wholesome articles of diet which they consider to be funny food. Some-times it may be possible to change people's minds. For example, Englishmen once considered the Frenchmen's taste for frogs' legs and snails disgusting. Today, these items are recognized as being particularly tasty. The main factor restricting their wider consumption is price. But about dogs and perhaps rats and mice as well, the British and much of the technologically advanced world are adamant. They refuse to eat them regardless of their nutritional excellence. Yet dogs have been eaten and done people good all over the world (with the exception, of course, of those areas where they do not exist). In the rain forests and savannas of West Africa and what was once the Congo not only were they eaten but many people preferred them to almost any other source of meat. The Tallensi of Ghana and the Poto of the Congo were pronounced dog-flesh lovers. So fond were the Mittu of the southern Sudan of dogs as food that in the old days of the slave trade people in the business used to breed dogs as particularly tempting commodities to pay for the slaves they bought. And at one time the Koma, who live in south Ethiopia, had a charming custom at the annual celebration of the new moon killing and cooking a dog and presenting its tail to their ruler to eat. As soon as he had consumed this delicacy his term of office was extended until the next celebration, in much the same way that the Lord Mayor of London invites the current Prime Minister to an annual banquet at the Guildhall. And then there are communities, in the Upper Congo, in Zambia, and also in Angola, where they used to fatten dogs for the table rather in the same way that we fatten chickens. In China, too, dog-eating was once very popular and the chow was actually bred specially for its culinary virtues. Nowadays, however, dogs have been abandoned as Chinese food.

For those readers who find these reports of historical truth distressing to their sensibilities, it can be pointed out that British aversion to the whole idea of dog-eating was illustrated in a particularly striking manner during World War II. At the time when Great Britain was beleagured and fighting alone against overwhelming odds, the estimate was made that even under the rigorous hardship of rationing for the 50 million or so people on the island, there were eating with them an almost equal number of pet dogs. If all the food being eaten by all these dogs (the calculation was made) was used to feed chickens it would be possible to issue every month an extra distribution of priority eggs to expectant mothers and in addition there could be a once-and-for-all bonus ration of sausages. Regardless of the hardship of the times, this tempting opportunity so clearly shown by the food scientists was unhesitatingly turned down. Science can provide all the useful information about protein that anyone might want. Whether such information is used or not is a matter to be decided, not by science, but by the feelings and aims of everyone – that is, by the community.

So we don't eat dogs! What then about dormice?

Dormice are not mice at all but are distant relatives of the squirrels. They are to be found all over the place – in Europe, Russia, India, China. For people like ourselves who make no bones about eating cattle, sheep, pigs, the occasional goat, rabbits, deer, frogs (if we're French) and whales (if we're Norwegian or Japanese) there are many advantages in domesticating dormice for the table. Before we recoil at the idea of consuming mice, I should repeat that a dormouse is not a mouse at all. There are a number of varieties of various sizes. All of them, however, move about at night rather than by day and, in their natural habitat, live in trees. The idea of using them for food is not new. In fact, the Romans were particularly fond of them and the kind of dormouse they ate is still well known in several parts of Europe as the 'edible dormouse', the scientific name for which is *Glis glis*. The way the Roman housewife maintained a supply of dormice for the table was to keep them to start with in nesting boxes. A pair of dormice usually produces four baby dormice twice a year, that is, a crop of eight. These grow quickly and can be kept, each one by itself, in an earthenware container and fed on kitchen scraps. Full grown they are about eight inches long.

It should perhaps be pointed out for the benefit of any reader who is considering raising dormice for meat that they need to be handled with circumspection, bearing in mind that Canadian beavers are distant relatives with whom they share the possession of quite long and slightly protruding front teeth. Consequently when they bite the hand that feeds them – as they tend to do when they are picked up – they can easily draw blood even through a pair of gloves. A special culinary advantage they possess arises from the fact that they are hibernating creatures. Before retiring for what they imagine to be the winter they become very fat and it is at such times that they are particularly esteemed as a favourite article of diet in a number of parts of the European continent. At the time of writing it must perhaps be accepted that for many readers the edible dormouse, nutritionally valuable and agreeable and pleasant to the palate though it may be, must nevertheless be categorized as a 'funny food' about which the more conventionally minded diner might feel some doubts. Surely, however, with the gourmet Romans as guides from the past and much of Europe as indicators for the present, the time may not be long delayed before dormouse on toast takes its place on the menu.

When people talk about foods of one sort or another – purple chipped potatoes or green bread, for example, have been tried out in scientific experiments – as being 'funny', they usually mean 'funny peculiar' rather than 'funny ha-ha' and use the word as an expression of disapproval. There is often no good reason for such dislike or, if there is a reason, it is buried somewhere deep in the history of the tribe. For example, the British think it disgusting to eat horse-meat and to tell a restaurant owner that the meat he serves tastes like horse is to insult him – in Britain, that is, although not necessarily in France. Nor does this mean that the complainer does not like the flavour of horse. He may never have tasted it. If he did, he would probably like it; there is a sweetness about horse fillets that beef or lamb does not possess. But though most foods that are described as 'funny' will be avoided by the people who so describe them, there are others that will be particularly esteemed. Oysters, for example, have for some time had a mystical smack about them and are believed to be particularly valuable to newly married young men. Tomatoes, too, when they were only recently introduced from America and were still strange to consumers in Europe, were also believed to be of assistance to lovers. It is perhaps less surprising, in view of the downright approach to such things of primitive societies – many of whose beliefs have come down to ourselves, whether we know it or not – to find that there are communities where the men recommend the bridegroom to eat bananas or leeks on his wedding night.

There are at least three groups in South Africa – the Tembo, the Fingo and the Nguni – who believe that when women eat eggs they become so eager for love that they chase after men even if they belong to a different tribe. Indeed, if a lady says to a gentleman, 'I shall cook eggs for you,' it means only one thing. In most of the areas where this belief is common it is not surprising to find that women are forbidden to eat eggs.

If we come back from remote and distant people such as the Tembo, the Fingo and the Nguni, we can find people much nearer home with funny ideas about the foods which either they do or they don't like. When chocolate was still a comparative rarity in Great Britain people quite soon discovered that they liked the taste. The next thing they decided, however, was that it was good for loving. This idea is, of course, pretty well exploded by now even among those who like a cup of cocoa last thing. But the extraordinary degree to which people cling to ideas about their likes and dislikes about food, even when it is obvious that they have nothing to do with nourishment, is remarkable. And the lengths to which scientists will go to provide them with what they want is even more surprising. Consider the bubbles in bread.

People today have come to expect that when they buy a loaf of bread each slice will be evenly dotted with holes (each one of which started life as a bubble in a mass of sticky dough) and that each hole will be, within very close limits, the same size as all the others. This is quite difficult for the baker to do. If he uses flour made from English wheat, not only will the bubble holes be different sizes but, in general, the bubbles will be smaller. This means that although there will be exactly the same amount of flour as before, the loaf (while being the same weight as the customer is accustomed to) will not stand so tall on the kitchen table. Who would think that this would matter? After all, take a mouthful of a taller slice cut from a loaf with bigger bubbles, or one from a shorter, lower loaf and chew them up and in 15 seconds no one could tell the difference. But it does matter. Bakers whose loaves don't stand up tall in their baking tins soon find they have fewer customers (that is, fewer people like you and me) than bakers with bigger bubbles. Naturally, it follows that the bakers get hold of the scientists and ask them 'Why do some kinds of flour produce bigger loaves of bread than others?' The answer is that the flour with the bigger bubbles has more protein in it and protein with a slightly different chemistry. No one at this stage is interested to know whether one protein is better for you than the other (as a matter of fact there is no difference). What they want to know is whether, when the flour is made into dough and bubbles form in it as it rises, the protein will blow up into a tough little balloon that will not burst.

There are probably more serious-minded scientists sitting in laboratories all over the world doing research on flour made from mixtures of 'hard' wheat from Canada, which can be bought in the shops as baking flour and compared with flour made from 'soft' wheat grown in England or on the Continent, which is sold as biscuit or pastry flour, than there are scientists studying nutrition. And all because people would think it 'funny' to eat a low-slung flattish loaf rather than a tall, bold upstanding one. If you don't believe me, try making bread with biscuit flour.

It used to be said that, as Anita Loos put it, 'Gentlemen prefer blondes,' that is to say, men like girls with yellow hair. And so many of them do, that quite a lot of girls go to the trouble of dying their brown hair yellow. This is regardless of the fact that they have just as sweet natures and cook just as well whatever colour their hair may be, while at night, with the lights out, who can tell the difference? The French say quite rightly that 'At night all cats are grey.' Nevertheless with food, colour, taste, smell, touch – and even sound – though they add nothing to nourishment, do exert an effect. Yes, even sound.

In the Swedish food-research laboratories there is a unit where scientists spend most of their time measuring the sounds food make when they are being crunched up. The experiment is easy to do. A crisp, rigid ginger biscuit which breaks with a loud snap when it is bitten is popular – there is nothing funny about it. Keep the same biscuit for a day or two in a moist atmosphere and its taste is exactly the same but nobody would eat it. It does not sound right; it is faint and out of key. The same can be said for damp cornflakes, and even with dry ones the music only lasts a minute or two after the milk has been poured on. Celery is another example of a food eaten primarily for its music. In actual fact, a stick of celery provides precious little nourishment. When eaten with biscuits and cheese, its primary function is to supply the sound accompaniment.

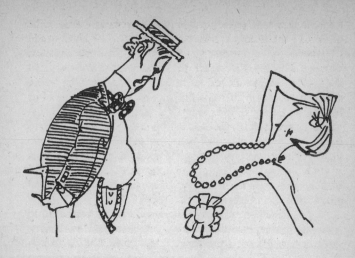

3 Zip goes the enzyme

Not so long ago there were no such things as zip-fasteners. Coats and dresses, trousers and skirts, were kept closed with buttons and buttonholes or with hooks and eyes just as they had been for thousands of years. There is no mystery how these work: it is obvious for everyone to see. But how does a zip work? And how did anyone ever come to invent zips?

The answer to the second question is that no one did invent zips right off, at least no one invented them in the present simple, neat and useful form. But there was a man, an American, Whitcomb L. Judson, who in 1891 had an idea that *led to* the invention of zip-fasteners. At that time it was fashionable for ladies to wear high boots which were done up on the outside of each leg by a series of hooks and eyes. It was quite a laborious business to fasten each boot; it could take a longish time and was hard on the fingers. Judson was struck by the notion that, instead of hooking up each hook into its opposite eye one by one, it would be a popular convenience if he could devise a mechanism that would do this in one operation. And so he designed a metal contraption, which when it was pulled up slid over the hook on one side and the eye on the other and then caused the hook to tip over so that its point fitted into the metal ring of the eye. No one had ever tried to do such a thing before so it is hardly surprising that his patent was the only one of its kind when he applied for it in 1891. Actually, his device did not work particularly well and he immediately set to work to improve it. And even by 1905, when he took out his fifth patent, he was still the only person who seemed to think it worth while to try to devise a mechanical slide to hook up ladies' boots. By then he had become more ambitious and he was trying to adapt his device to hook up ladies' dresses as well.

At this time, not only was no one but Judson interested in *trying* to do what he was attempting but he himself was finding the task difficult and frustrating. It is a curious thing that so many of the actions that are quite easy to do with one's hands and feet prove to be exceedingly difficult to mechanize. For example, several research workers in the 1970s, interested in what is called *bioengineering*, have attempted to design a machine capable of walking upstairs and have found the exercise so complicated as to be virtually impracticable.

Poor Judson had endless trouble with his device for hooking together hooks and eyes. Even when a friend of his, Colonel Lewis Walker (who knew him because of their joint interest in one of the early private tram systems in Chicago, the Judson Pneumatic Street Railway Company), organized another company to make the newly invented hook-and-eye doer-upper, his difficulties were not over. The firm was called the Universal Fastener Company and to start with employed a lot of girls to fix by hand rows of hooks and eyes on strips of material so arranged that when the slide was pulled up the two strips became locked together. They tried to devise a machine to do away with the laborious business of fixing the hooks and eyes by hand but the first one made was so complicated that they found it easier to go back to the girls. In the end Judson modified the whole idea and instead of fixing one hook above the other into a chain of eyes similarly fixed together, he hit on the idea of clamping the hooks and eyes to the edge of lengths of tape. This system made it easier to manufacture the whole thing on a machine.

By the time he had got so far, the enterprising Mr Judson reorganized his company and called it the Automatic Hook and Eye Company of Hoboken. He also arranged for pedlars to go round the country selling his automatic hooks and eyes. But although his device had achieved some success for zipping up (though it was not called this yet) ladies' high boots, it was not entirely successful when applied to garments because it had a tendency to spring undone at unexpected moments when put under strain.

From 1905 to 1912 Judson's company struggled on, on the verge of bankruptcy, but even though he and his collaborators changed the design a number of times they never seemed able to solve the problem of how to stop the hooks coming un-hooked, even when they had designed a pretty good slide to get them to hook up. Indeed, the whole project was on the point of collapse. It is interesting to reflect that if this had happened, zip-fasteners would not have been invented. By good luck, how-ever, the Automatic Hook and Eye Company at about this time took on a Swedish electrical engineer called Gideon Sundback who knew nothing about zip-fasteners – of course, nobody did. Sundback had actually been working for the West-inghouse Company ever since he arrived in the United States. He was an inventive person and, after making a number of modifications, in 1913 he at last designed a fastener which was in all essential principles the modern zip-fastener we know today.

Sundback's contributions to this success were of two sorts. First of all he made a series of step-by-step improvements. To start with he built up the zips on lengths of tape so that the whole business could be fixed on to the garments for which they were intended. And he made the fasteners more flexible, un-like Judson's early models which were mainly constructed of metal links. But Sundback's genius lay in getting away entirely from the hook-and-eye principle. This involved a major intel-lectual jump since Judson's notion was essentially a device for doing up hooks and eyes. The modern system which came out of Sundback's head does not call for the point of a hook to be brought down into the hole of an eye. There are neither hooks nor eyes. On each side are moulded teeth which are brought together by the two sides of the slide so that, as they mesh, a hump in the top of the first fits up into a hollow in the bottom of the second and a corresponding hump in the upper side of the second fits correspondingly into a hollow in the underside of the third and above it, and so on. Not only is the exact shape of the teeth important but so also is the precise design of the slide which brings them towards each other and fits them together. The interesting point in the story is that although Judson must be given the credit for the original idea of something like a zip-fastener which was to emerge from a system of doing up a row of hooks and eyes automatically, only when Sundback gave up the idea of hooks and eyes and conceived the notion of moulded teeth fitting together one above the other was it actually found possible to construct the first zip-fastener. Sundback not only invented a practical fastener but he also designed machines for stamping out the parts and for fixing them on to tape.

As a matter of fact, even after zips were invented and on the market, it still took a lot of effort to persuade manufacturers to use them. Despite their obvious convenience, clothing manufacturers still clung tenaciously to buttons and buttonholes and to hooks and eyes. The real advance came only in 1918 after five years' hard slog. And as with so many other innovations, the impetus came from the war. A manufacturer making flying suits for the American Navy decided that to keep the wind off airmen exposed in the open cockpits of those days he would fit the garments with zips. He purchased 10,000. Their future was assured after the war was over when the Goodrich Company decided to put zip-fasteners on their galoshes.

This is the answer to the second question which was asked at the beginning of the chapter: 'How did anyone ever come to invent zips?' The first question 'How does a zip work?' has perhaps not yet been properly answered. Part of the answer is that a zip does *not* work like a system of hooks and eyes, as Judson thought it should. In such a system, each hook and eye hold together quite independently of any other hook and eye. With a zip, the two sides only hold together so long as the whole rank of all the teeth on one side stay meshed into the whole rank of teeth on the other side. That is to say that whereas each hook, or each button in a system of buttons and buttonholes, works independently of all the others, with a zip, after the slide has pulled them into each other, all the teeth must work together or else none of them is any good. Everyone has experienced the disaster of a broken zip; once a bit of it breaks undone, the whole thing flops open.

The real answer to the question 'How does a zip work?' is that it works because there are two sides that fit together as naturally as one's two hands do when one clasps one's fingers. If a wrestler decides to interlock his fingers with those of his opponent in an 'interlock', there is almost no way he can get free unless either he is able to break the hold by sheer brute force – and the amount of force required is considerable – or he

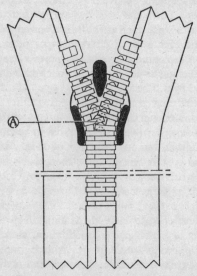

The illustration shows how the protuberance on the upper side of each tooth fits into the indentation in the underside of the opposite tooth above it (A). The slide brings them together as it is pulled up.

can persuade the other man to release him. When a zip is closed, all the 'fingers' on one side are interlocked with those on the other. The secret of how each set of 'fingers' is designed is no more subtle in its way than that of how the knuckles of the fingers in the wrestler's 'double-handed lock' fit up against those of his opponent. Where the zip does contain a remarkable piece of design is in the exact construction of the slide by which it is done up and undone.

The slide of a zip must possess three vitally important characteristics. Firstly, it must gather up the teeth from the two sides so that those on one side fold over and interlock one after the other with those from the other side. But the most important quality the slide must possess is that it should be *exactly* fitted to the size and shape of the two rows of teeth. Zip-fasteners, now that their usefulness and versatility have been recognized, come in all shapes and sizes. There are big coarse ones on suitcases and tiny delicate ones on the most intimate of garments. There are zips fitted to rubber galoshes, plastic document cases and woollen jerseys. There are zips in which the teeth are made of metal and others in which delicate flexible teeth are so fine as to be barely visible at all. But whatever their design, size and structure each one possesses this essential property: the slide that opens and closes it must be the right one and must fit as accurately as a key fits in a lock.

It may be said – and people would almost be right to say so – that zip-fasteners have nothing to do with science. Nobody knows *why* Judson was suddenly struck with the idea of doing up hooks and eyes automatically. Although it is a nuisance, it is not particularly difficult to hook up a pair of ladies' high boots. And it might be considered too large an investment of time and effort to work from 1893 to 1923, a period of thirty years, to perfect a device just to do so straightforward an operation. But science is a strange business. The scientist is trying to discover the secrets of nature. But to do so he must have tools. Nobody could do very much about infectious diseases until someone else had invented a microscope to allow the bacteria (germs, if you like) that are the active cause of many diseases to be seen. In our own time, it would be impossible to make a fuss about pollution if chemists had not invented ways of detecting whatever it is that is causing the pollution.

This same marriage between an invention – a zip-fastener this time – and the basic understanding of how things happen seems to be turning up now that zips have become so varied, smooth-running and efficient. When you chew up and swallow a piece of potato or a mouthful of bread, much of the stuff that lands up first in your stomach and later in your intestines is starch. If you looked at starch with the best optical microscope you could get, then with an electron microscope and, besides this, made use of all the scientific evidence there is about starch, it would gradually become clear that starch itself is made up of a tangled web of glucose units meshed together rather like knitting. Starch as such, that is, this web or mat of glucose 'stitches', cannot be digested through the lining of one's intestines. What happens is that during the process of digestion the glucose 'stitches' are unfastened from each other and, being much smaller than the lump of starch they come from, they *can* be absorbed.

This is a short account of what happens, but *how* it happens illustrates what an elegant and peculiar system the living body is. I have just said that in chemical terms starch is made up of long chains of glucose units. It can now be shown that each of the glucose units which, like beads in a necklace, is linked to the one next to it, is joined in such a way that they are all the same way up. The glucose links are made up of a little ring of five carbon atoms and an oxygen atom with a protruding piece made up of one further atom of carbon. The whole thing is hung round with hydrogen atoms and more oxygen but in the main it is this ring of atoms with one carbon atom hanging out. And when I say that in starch all these links are the same way up, I mean that the odd carbon pieces all point in the same direction.

Under normal circumstances, starch is a pretty tough material. A woman ironing a shirt can spray starch on it and run the hot iron over it without destroying the starch. Starch stands up to the temperature of the oven when bread is being baked and even when the bread is toasted the starch does not fall to bits even though it breaks down into *dextrine* when the toast becomes brown – I say nothing about the black ashes of burnt toast. How then does starch so quickly unravel into sugars (glucose and maltose are classified in scientific terms as 'sugars') when one eats it? The answer is that in the digestive juices there is to be found an *enzyme*. The best way to explain what this is is to compare it to the slide of a zip-fastener, because that is what it amounts to. An enzyme is a natural substance that is made in the body exactly to fit – as precisely as a key fits a lock – the particular job in the chemistry of the body that it has to do. One kind of enzyme (like one model of zip-fastener slide) will only do one job. The one that works on starch in the intestine *only* unravels starch and nothing else. But this it does smoothly and quickly, which is why nobody has any trouble digesting a slice of bread or a biscuit with their early morning cup of tea.

On the other hand, nobody can digest sawdust or paper. Yet in some respects it is odd that they cannot because both wood and paper are largely composed of cellulose which, like starch, is made up of chains of glucose links fastened together. But although both starch that we can eat and sawdust and newspaper that we cannot are made up of chains of glucose, there is a difference between them. Whereas the glucose links in the starch are, as I have already said, all connected together the same way up, the links in cellulose are connected differently: the first has its carbon stalk upwards, the next has its projecting carbon downwards, the next upwards, the next downwards, and so on all along the chain. It is natural, therefore, that the enzyme that unzips the glucose units in starch, like a slide designed to fit its own particular zip, won't work on cellulose which is, as it were, a close zip needing a slide specially designed for it and quite different from the one that fits on to starch.

It is astonishing how closely the zip-fastener principle illuminates the subtle way the body works. At every point in the body's mechanism there are enzymes exactly suited for doing what they do. For instance, we get the energy that makes the body work by further breakdown inside us of the glucose I have been talking about after it has passed on into the body. At one stage in the process there is one particular enzyme which, like all the rest, is exactly designed to release a bit of the

47

glucose that has already been partly split up. But it does not work as it is: it is like a zipper slide that has lost the tag that is needed to pull it along. The extra bit it needs is called, in the jargon the scientists use, a *coenzyme*. The coenzyme on its own is not of much use, any more than a tag – that is, the part one gets hold of – is when it gets pulled off the slide.

This coenzyme that combines with the enzyme that pulls glucose to pieces and lets its energy out is a combination of one of the B-vitamins one gets from such foods as bread and bacon and, in fact, quite a variety of other things. But when people do not obtain as much of this vitamin as they need, which is when they are poor and trying to live on white rice and little else, the enzyme we have been talking about, lacking its coenzyme (the tag on the zip, as it were) does not work properly and the half-broken-down glucose (like a jammed zip-fastener) becomes stuck. Not only are the people not able to get the energy they need out of their food but the half-broken-down glucose piles up in their blood stream. This half-unzipped stuff is called *pyruvic acid* and the condition due to the inability to unzip is the disease, *beri beri*. When people die from it, they are dying from pyruvic acid poisoning.

4 The pop, pop, pop explosion

A good way not to explain to a learner who wants to know how a motor cars works is to say that the pistons are kept going by a series of explosions. Ordinary people have in their minds a clear idea of what an explosion is, and it certainly is *not* the gentle purr of the well-tuned engine of an up-to-date family car. A car is supposed to go smoothly along, and many of them do. People's understanding of an explosion – as a shotgun blast, as the variety of firework known as a thunder-flash, as a paper bag bursting, or even as a car backfiring – is very different from the gentle progression of traffic down the high street. A large, so-called 'juggernaut' lorry is noisy, to be sure, but the noise it makes is not what is popularly meant by an explosion. Yet the popular interpretation may be wrong or, if it is not wrong (scientists, after all, have no right to remodel the English language), it is in some respects misleading. By being restricted to a noisy and sudden release of energy, the term explosion obscures the significance and meaning of a whole lot of interesting phenomena.

Men who have charge of flour mills know a good deal about the material they handle. Just the same, it is doubtful whether a miller would have much to say about the nourishment to be had from eating products made from flour and, even if he did, he would be unlikely to say a great deal about its energy value in terms of calories, which nowadays are called kilojoules anyway. All competent millers, however, know very well that in their trade flour can be a dangerous material: they know all about the hazards of dust explosions and take special precautions to guard against their happening. Yet the dust explosion in a flour mill and the energy to be obtained by eating a slice of toast possess many properties in common. Let me first of all illustrate the nature of a dust explosion in which the dust is flour.

You first take a tin can (a cocoa-tin will do), push a rubber tube through a hole drilled in the bottom to receive it, put a funnel inside the tin with a small amount of very dry flour in it and fix the stem of the funnel into the rubber tube. Then you stand a small piece of burning candle in the bottom of the tin, put on the lid and straight away blow into the rubber tube. If you have set up your experiment correctly – and you must have a fine orifice at the end of the tube joined to the funnel – there will instantly be a violent explosion and the lid of the tin will blow off. (Take care no one is standing near the tin.) The explanation of this dramatic experiment is that flour, the main component of which is starch, contains energy. This, of course, is well known to every lady (and a good number of men as well) who is interested in slimming, although they may not have realized that this could have much to do with explosions. Put into scientific terms, the flour can be said to be a fuel, and the fuel in it is combustible carbon and some combustible hydrogen as well. If the scientist happened to be a chemist, he might say that the carbon and the hydrogen are present in starch in the 'reduced' state. All that this means is that when they combine with oxygen the chemical energy they contain is released. What happens in the cocoa-tin bomb is that the flour is caused to fly up as dust when you blow into the funnel, the heat of the candle flame makes it combust very quickly with the oxygen of the air blown in and the sudden release of energy goes bang and blows off the lid. In a flour mill, if by chance dust is allowed to rise and there is a flame about– or even a lighted cigarette or a spark struck by a stone that happens to have got between the steel wheels of the milling machinery–the explosion can be devastating.

On the other hand, put a candle flame to a pile of flour and nothing much will happen. At worst, the flour may become charred; if it burns at all, it only does so reluctantly. A spark falling on such a pile is more likely to go out than anything else. This does not mean, however, that heaped flour contains any less chemical energy in it. The cloud of flour dust in the tin explodes with a ringing bang because as a cloud of dust each tiny particle of flour is surrounded by oxygen from the air. This pile of flour, or at least the outside layer of it, is in contact with air, to be sure, but there is less of it in proportion. Furthermore, the energy released as heat and as the products of combustion from the small amount of flour that does become charred is not reinforced by the combustion of more flour. The flour inside the pile has no air up against it to combust with, so it does not burn at all. This explains why it is that flour is not, in general terms, considered to be an explosive.

For ages the best explosive known – indeed, the only explosive known – was gunpowder. What, you may ask, is so special about gunpowder and why do we esteem the Chinese as being so sophisticated and scholarly in having invented it several thousand years ago? The particular characteristic of gunpowder is that it is composed of a mixture of substances full of chemical energy (this in ordinary terms usually means things that will burn) with a substance rich in oxygen that eliminates the need for oxygen from the outside air. All that needs to be done, therefore, is to mix the substance supplying the oxygen very intimately with the energy-rich substances which are to do the burning. The substances in the Chinese mixture we call gunpowder – that the civilized Chinese used for fireworks celebrations and that we others adopted so enthusiastically for warfare – were charcoal and sulphur to burn and saltpetre to make them burn.

But just to mix charcoal, sulphur and saltpetre does not

necessarily make good gunpowder. To start with, no two mixtures may behave alike and some of them may not even go off at all. To make good gunpowder – like doing good cooking – the ingredients must be carefully sifted so that all the granules of which they are composed are of uniform size. This allows the gunpowder-maker to mix them in a uniform way and it also allows the different ingredients to be brought into intimate contact with each other. During the course of its long history, gunpowder manufacturing – like cooking – was an art rather than a science. Apprentices learnt how to make it from their masters. Tricks of the trade were worked out; for example, it was found that if the mixture, when its ingredients had been carefully amalgamated together, could finally be produced in the form of fine pellets rather like lead shot, the explosion, when the gunpowder was set off by a match, travelled more quickly through the mass of powder and the force produced was thereby increased.

Gunpowder, simple though it is, was not particularly satisfactory. To start with, it produced a lot of smoke. Five minutes after the start of a lively battle, the battle-field was masked in blinding fog. Furthermore, gunpowder does not release so very much energy. It followed, therefore, that cannon balls and shot did not really go very far. All the same, a general who could obtain not only all the gunpowder he needed but *good* gunpowder did have an advantage over an opposing general whose gunpowder was of inferior quality. This was largely a matter of obtaining the ingredients.

Charcoal, which is carbon, was probably least trouble to obtain. It was produced by charcoal burning, an ancient occupation, from the wood of the forest. Sulphur turns up naturally here and there and has been known for ages. Hell, for example, was often depicted as being full of burning brimstone (the old-fashioned name for sulphur) where the damned souls, apart from being burned, also suffered the inconvenience of coughing and having to blink their smarting eyes. It is when carbon and sulphur combine with oxygen that the energy in them is released. The black granules of carbon turn into oxides of carbon, the most common of which is the gas, carbon dioxide, which we ourselves breathe out as the 'ashes' of the

'. . . Why do we esteem the Chinese as being so sophisticated and
scholarly?'

carbon we have oxidized inside ourselves in getting the energy out of our food. The sulphur is converted into oxides of sulphur which are also gases.

Saltpetre is, in chemical terms, called potassium nitrate. One does not need to be an intellectual genius to deduce that its name implies it to be a compound of nitrogen. One would, however, need to know at least a certain amount of the way chemists make up their technical words to apprehend from the *ate* part of its name that it carries in its substance a proportion of get-at-able oxygen.

Where does saltpetre come from? This, for a long time after gunpowder had been introduced as the best and most destructive agent for killing people and knocking down their castles, was a question of urgent importance. It was – and still is – found here and there as a deposit on the surface soil in India and it appears likewise in parts of Spain. But what were the nations who possessed no natural deposits to do? This is a question of current importance today except that nations are now asking themselves how to obtain energy to produce the controlled purring explosions of petroleum in their mechanical engines. It would be romantic indeed if the solution found for the gunpowder problems of our ancestors were applicable to the petroleum problem of ourselves.

The saltpetre-poor (but warlike) nations of Europe turned their attention to the exploitation of the incrustations of saltpetre which were known to occur on the damp, dirty and smelly walls of stables and old cellars. Men were sometimes employed to ferret about looking for dirty old cellars and smelly old stables with worthwhile layers of incrustation on them. Naturally, it was not long before the reason was discovered why some cellars and stable walls were more deeply incrusted than others: more manure and muck had been left there – and left longer. It was not long before governments did something about increasing the yield of this military supply by artificial means.

Certain areas outside towns or villages were designated as 'nitre beds'. Here manure and garbage, vegetable refuse and the like were piled up to a height of three or four feet on floors made of clay, or sometimes wood. The whole thing was covered over by a roof and periodically moistened either with slaughter-house blood or urine. The national welfare has been a hard taskmaster throughout the ages and the production of saltpetre must have been a sore trial for the local inhabitants – particularly such of them as lived downwind – when it is recalled that the nitre beds were kept going for about two years before they were considered to be properly mature and full of saltpetre. Napoleon, being an efficient administrator, was especially systematic in the steps he took to ensure adequate supplies of saltpetre: besides encouraging the setting up of nitre beds he issued a number of ordinances insisting that the citizens should urinate on them in appropriate amount.

Arrangements enforced in Prussia were rather different. There, the farmers were compelled to build up and maintain piles of rotting organic matter as walls around their fields. These walls were pulled down and taken away by the public authorities when they were 'ripe'. In Sweden another arrangement again was in force: the country people were asked to pay part of their taxes in the form of decayed and saltpetre-rich 'compost'.

Early on in the thirteenth century, gunpowder was a mixture of about 30 per cent charcoal, 30 per cent sulphur and 40 per cent saltpetre to set the other two off. This did not make a particularly good explosive. Now that the chemistry of what happens is much better understood than it was 600 years ago, it is clear that the ideal mixture is approximately 13 per cent charcoal, 12 per cent sulphur and 75 per cent hardly won saltpetre. In the early days, and particularly when the three ingredients were not properly purified, not only was the formula far from ideal and the gunpowder that much weaker than it had need to be but there were also times when it did not go off at all.

The history of gunpowder is worth pursuing further. To start with, the gunpowder people mixed up their ingredients and then laboriously ground the mixture into a fine powder. This was very difficult to use and required considerable skill on the part of the gunner. If the powder was packed into the gun too tightly, when the time came to fire and a match was put to the touch-hole, either the mass refused to burn at all or it burned quite slowly and exhibited hardly any explosive effect. On the other hand, if the gun was charged too loosely, the powder could go off abruptly, a significant proportion of the gas and energy would escape as 'windage' round the sides of the shot, and a short-ranged ineffective shot result. This brings us to an important point that lies at the heart of the matter: provided the gun is loaded properly, the finer the mixture, the more powerful the final explosion will be.

But then a new matter arose. As guns became larger and their barrels were rifled to give the shell a twist so that it remained on course when fired, the amount of work required to get a big shell from A to B became very substantial. So much so that when the necessary amount of finely powdered gunpowder was put in to send the shell to where it was intended to go (and every general wanted his cannon ball to travel further than the other fellow's), it tended to burst the gun barrel. It was then recognized that not only was the amount of force obtainable from any kind of gunpowder important, but so was the speed at which the force was delivered. Several ingenious developments were invented. As early as the fifteenth century it was spotted that if a fine mixture of sulphur, charcoal and saltpetre was moistened – and the most popular moistening agents were alcohol or (and this was esteemed to be the ideal fluid) the urine of a wine drinker – it could then be beaten into a cake, re-dried, powdered up again and sifted. Large-grained gunpowder was found to burn more slowly than the fine powder, and to be more useful as a 'propellant', as it got the shell into motion up the barrel of the cannon more gradually without bursting the barrel.

Towards the end of its history all sorts of ingenious ideas were introduced into the improvement of gunpowder. But all the time it was gunpowder that was being improved and all the time it was the chemical energy in the 'fuels', charcoal and sulphur, and the oxygen supplied by the saltpetre that allowed the explosion to take place. By the use of fine powders in muskets and shotguns a short, sharp explosion was achieved because the gunpowder burned quickly. In cannons and larger ordnance, bigger granules of powder were used to produce a slower, shoving push up the barrel. Such improvements were the results of some 600 years of study, true, but not of particularly scientific study. Perhaps the effort can better be described as trial and error.

Then suddenly – or, at least, suddenly in historical terms – science was applied instead of trial and error. People began to ask themselves why they needed to mix two or three substances together? Might it not be possible instead to produce a single chemical compound (not a mixture) which contained both carbon and hydrogen as fuel in one end of the molecule and oxygen or an oxygen-giving compound at the other? If this could be done, surely the explosive would go off more reliably, more quickly and with more force?

It was in 1845 that a German called C. F. Schönbein brought in a new era with the discovery of guncotton or, in more exact terms, nitro-cellulose. He made it by treating cotton (which is mainly cellulose) with nitric acid. Guncotton is indeed explosive. So much so that for twenty years after its discovery a whole series of people were killed by being blown up while trying to handle it. The 'nitro' part was all too intimate with the cellulose part.

An Italian, Professor Sobrero, invented nitro-glycerine at about the same time that Schönbein discovered guncotton. Glycerine is the pleasant, sweet liquid sometimes used in cough mixture. When it is treated with nitric acid a quantity of nitro-glycerine is obtained which is so likely to go off that no one in his senses would have anything to do with it. The molecule is so tense and strung up that it only needs someone to drop the container – nay, even to jar it on the table may be enough – and the single molecule of nitro-glycerine splits up into its parts, re-arranges itself, gives off heat in doing so and thus – explodes.

One of the great romances of modern history is indissolubly linked with the name of Alfred Nobel. Everyone has heard of the Nobel Prizes, large money prizes given not only as the highest distinction for science or literature but also for contributions to peace. Yet Alfred Nobel acquired the huge fortune from which the money for the Nobel Prizes is derived by inventing dynamite. He started by trying to market nitro-glycerine, which he put on sale as 'blasting oil'. The awful dangers of handling this stuff were brought home to him when in 1864 his own brother, who was also in the business, was killed in an explosion. Three years later he discovered that if the nitro-glycerine were soaked up in a type of porous clay called kieselguhr and moulded like sticks of plasticine, it could then be handled with tolerable safety by people who knew what they were doing.

But equally important were the later discoveries of how to make other chemical molecules for specific explosive purposes: both slow-acting explosives (to act as propellants and push shells out of guns without bursting the barrels) and also quick-acting explosives, which might either be used as detonators, to set off the slower explosives, or as bursting charges in shells to blow up buildings and kill people. In no time there was *fulminate of mercury* (the stuff that is put into the cap at the end of a rifle bullet or shotgun cartridge to explode the main charge), cordite, lyddite and – the most versatile killer of them all – TNT.

To sum up; an explosive is merely a substance capable of exerting sudden pressure on its surroundings. Obviously some explosives are more destructive than others. A pound of TNT goes off with a bigger bang and releases more energy (all of which is wrapped up in its own molecules to start with) than a pound of gunpowder (in which the energy is derived from the three ingredients). And a pound of gunpowder has more destructive force than a pound of flour which needs to get hold of enough air to combust with at the critical moment. But flour is an explosive nevertheless.

Let us move one stage further. Every now and then one reads in the paper of explosions taking place in sewers. What is it that explodes with force enough to throw cast-iron manhole covers into the air and kill people? The world is full of a wide variety of creatures all doing their best to live their own lives. In the jungle of the world, cows and sheep eat grass and corn and themselves get eaten by lions and tigers – if they happen to live in those parts of the world where lions and tigers live – or by us. When we eat food we do not use up the whole of everything to keep ourselves going. Some of it passes right through our digestive system and some of it, like the urine I have been writing about, still contains a residue of chemical energy even when we have done with it. It is only to be expected therefore, that, just as the main purpose of the pigs and dogs our ancestors kept was to eat up the skin, bones, scraps and offal they wasted in their rubbish piles, so also is there a whole zoo of micro-organisms waiting to utilize what we waste. These beasties (as Robert Hooke, an excellent scientist, called them 200 years ago) get at the energy in our wastes as best they can. Sometimes they themselves have to let some of the chemical energy escape too – often in the form of gases such as hydrogen and methane (which is what coalminers call 'fire damp').

Left to herself, a cow could get no more nourishment than a man out of straw. But she is not left to herself. In her rumen, which is the large bag in which she stores her cud, there is a whole teeming mass of micro-organisms. But these creatures are not there to get rid of the cow's food: on the contrary, their purpose, so far as the cow is concerned, is to provide her with extra food by making much of the straw digestible. In doing this, they break up the straw's chemistry, getting their own food out of it, but also leaving bits behind uneaten. And among these chemical residues, besides what the cow can now digest, are the gases, hydrogen and methane.

I once knew a man who was doing research on the various kinds of creatures that are to be found in a cow's rumen. This is an interesting, complicated and difficult job. Even today there is still more exploration to be done and unknown organisms to be tracked down. As a means of studying these creatures, my friend, who was an accomplished vet, had constructed an opening in the cow's side. This sort of thing is often done by surgeons on people and is called a fistula. A cow's fistula tends to be bigger than a person's, and so it was with the fistula in my friend's cow. Except when he was taking samples, my friend kept his cow's fistula closed with one of those snap-on caps one finds on a five-gallon petrol tin. The cow did not seem at all put out by this arrangement. But as she quietly ate her food, I noticed that my friend had hung a notice on the cap on her side. It read: 'No smoking'.

We are, of course, back to explosives. Deprived of air,

whether in a cow's rumen, a sewer or, for that matter, in our own large bowel after we have been eating baked beans, the kinds of micro-organisms that thrive are those that produce hydrogen and methane, both of which, if given the chance, and provided with the right amount of oxygen, are capable of exploding, or of working an internal combustion engine – such as that in a sewage works (or in a motor car). In an up-to-date sewage works where the very same kinds of micro-organisms that may be found in a cow are put to work on a large industrial scale to break down the sludge we want to get rid of, methane and hydrogen are collected in gasholders and are used in machines not very different from those used in motor cars. The repeated throbbing of the machines, just like the throbbing of a well-tuned car engine, is due to a series of explosions.

In spite of what was written at the beginning of this chapter, what goes on inside a motor car engine *is* a series of explosions. Like anything else, explosions can come in various sizes; there may be big ones that blow down half a city or, on the other hand, small and controlled explosions that give a well-tuned engine flexibility. In fact, when it is defined in scientific terms, an explosion is merely an example of the rapid release of energy. The energy may start as *chemical* energy – when carbon and hydrogen, for example, combine with oxygen to produce carbon dioxide gas and water. Some of this energy may be in the form of heat. When this happens, the water will be steam and the thrust of this together with such other hot gases as may be produced supply the thrust of the explosion.

Let us, however, move one step further and ask, is it reasonable to speak about one man punching another on the nose with 'explosive' force, or of a man 'exploding' into action? Perhaps it is. If the body of a man or woman is thought of as a machine for moving about and doing things, it can be seen to be made up of a loosely hinged skeleton operated by a series of elastic cables. Years ago the experiment was done of cutting off a frog's leg, hanging it up and then putting an electric current through it from a battery. As the current flows, the elastic muscle in the leg contracts and the leg twitches. A similar thing happens when a doctor seats his patient, gets him to cross his legs, allowing the upper leg to dangle, and then taps him just below the knee-cap: a signal goes to the muscle stretched over the knee-joint and the leg kicks.

To punch a man on the nose all one needs to do is to 'wind up' the chemistry of the proper bunches of muscle fibres, pull the trigger by making up one's mind to punch, fire off the appropriate nerve signals and *bang*. The stored energy wound up into molecules of special and rather complex compounds that can hold 'explodable' fuel is released and – all together – the muscle fibres stretched over the outer side of the elbow-joint and the shoulder-joint abruptly shorten, the arm shoots out and (provided the whole thing has been properly aimed and the man whose nose is being punched has not done anything effective to get out of the way) the punch lands and down he goes. The *detailed* chemistry of what happens is obviously quite a complicated matter. In summary, however, the chemistry is simple and is remarkably similar to the chemistry of other explosives. Panting and flushed with success, all the puncher has done is to 'explode' the carbon and hydrogen and part of the nitrogen of the food he had previously stored away in his body, and the cloud of 'smoke' that drifts invisibly away as he stands triumphantly over his fallen opponent is the same carbon dioxide and H_2O (plus a few more complex gases and a good deal of heat) that would be found had the explosion been caused by gunpowder.

5 The bubble in the beer

Should you ever be marooned on a desert island with a sack of barley and feel in need of a glass of beer, the thing to do is to wet the barley, keeping it warm, but not too warm, and wait for it to sprout. Two things happen: the roots begin to grow and a shoot – what maltsters call the *acrospire* – starts to grow too. The shoot begins inside the husk. In a few days, however, the tip of it breaks through and one can see it peeping out at the end of the barley grain opposite the roots. When this happens, one should gently dry the barley, taking care not to allow it to get too hot. The dry, sprouted grains can now be called *malt*. Put a grain of malt between your teeth and bite it. It will break crisply and, when you chew it up, it will taste sweet. What has happened is that the starch has broken down chemically and turned into a type of sugar called *maltose*. If you grind the malt, brew it up in hot water and strain off the husk, you will be left with a brownish, sweet, sticky liquid. This is malt extract. The next step is to leave the liquid to ferment. This happens because yeast, which is almost always to be found floating about in the atmosphere, gets into malt extract and gets to work in it. If one is an ordinary brewer or a normal citizen on dry land, it is a good idea to get hold of some brewer's yeast from a brewery. But on a desert island, with a bit of luck, there will be yeasts there anyway. They will make beer for you, even if it does not have quite as good a flavour as you have been accustomed to.

What the yeast is doing is trying to live its own life. To do this it makes use of the sugar, into which the starch has been turned during the process of germination. In getting the goodness it needs out of this sugar, the yeast breaks it down into alcohol and, as a waste product, blows off carbon dioxide gas, much the same as we do. We blow off our carbon dioxide in our breath while the yeast lets off its carbon dioxide as bubbles.

Should the beer be put into a bottle or, when the main fermentation is over, into a keg, the carbon dioxide gas will not be able to escape but will be compelled to stay hidden in the beer. As anyone who has tried his hand at wine-making or home-brewing will know, there are sometimes occasions when the gas, struggling to get out, succeeds in its struggles and bursts the bottle or at least blows out the cork. For the most part, however, bottled beer remains still and quiet until the bottle is opened. When this happens, the dissolved gas has a chance to expand and as one pours out the beer, the bubbles come out too.

This is a general description of what happens. It is what we all know because we see it happening. And because we see it happening like this, not many of us become great scientists, nor do we make discoveries by which the nature of the world is better understood. What should make us ashamed of ourselves is that it is by looking at simple everyday objects like the bubbles in beer or the clouds rolling by in the sky that men and women who *are* scientists make their discoveries. And it was just such a scientist, D. Glazer, who in 1952 was gazing reflectively at his glass of beer. This is something that an ordinary man does hundreds of times in the course of a normal life yet, although he looks into his glass, he makes nothing of what he sees. Glazer, however, observed and noted what each one of us had often noticed without thinking, that the bubbles do not begin at random throughout the volume of the beer. They stream up in line, one behind the other. Look more closely and you can see that each column of bubbles starts at a tiny irregularity or point on the smooth surface of the bottom or side of the glass.

It is very easy to do a striking experiment to show what is happening. After a glass of beer has been poured and the first rush of bubbles has got away, there is still a lot of gas remaining in the beer (as can readily be checked by drinking it). Hold up the glass to let the light shine through it and there will be little gas to be seen: what remains can now stay in the beer. But should one provide the gas with a nucleus around which gas bubbles can form, form they will. For example, drop a single grain of sand into the beer and, as it sinks, bubbles will rise all around it and continue to do so even when it reaches the bottom of the glass. Better still, tip in a tablespoonful of sand and the beer will go mad. Where each grain of sand falls through the beer, bubbles will form and the beer – apart from being undrinkable, unless you like your beer full of sand – will froth right out of the glass.

What is the good, you may ask, of frothing, sandy beer? In actual fact, it is no good at all. But the *principle* by which bubbles form along the track of a moving particle can be very useful indeed. And this is the idea which was spotted, and used, by Glazer. If instead of using beer as the liquid one uses liquid hydrogen, the whole thing can form the basis of a *bubble chamber* which can be used to follow the track of charged particles emitted by the nucleus of the elements of matter. The bubble chamber is in fact the basic tool by which the science of nuclear physics has been elucidated. Nuclear energy, used in nuclear power stations and nuclear bombs alike, is derived from energetic particles of different sizes, possessing different amounts of energy.

Hydrogen becomes liquid when it is cooled to −247°C at atmospheric pressure. Put it another way, at this temperature and pressure, liquid hydrogen boils. If, however, it is compressed under a pressure of five atmospheres, it does not boil. That is to say, bubbles do not form in it any more than they do in a capped bottle of beer. But if the pressure is somewhat relaxed and at the same time a charged particle is fired into the

'. . . although he looks into his glass he makes nothing of what he sees.'

vessel, as it passes through the liquid hydrogen the particle will strip off electrons from the atoms of hydrogen in its path and leave a trail of broken atoms, called *ions*, to mark the track. These, like the grains of sand falling through beer, cause bubbles to form and enable the particle's track to be seen. Because of the speed with which events take place, such tracks cannot actually be 'seen'; they can, however, be photographed. Sometimes particles heavy enough and moving fast enough happen to collide head-on with a hydrogen atom and then it too may be set in motion. This effect is something like a fast-struck billiard ball hitting another, although sometimes in nuclear physics the struck 'billiard ball' is not only sent off on its own but may actually be broken into bits.

One of the great advances in human knowledge of the nature of the universe has taken place in our own times. It is the new understanding of atomic physics. It is interesting for me to look back at the physics I was taught as a schoolboy in 1925 and 1926. It is almost as if I were a Rip Van Winkle who, having been born and educated in a different scientific world altogether, had suddenly woken up in the modern world of nuclear energy, radio and television; it could almost be described as a new earth and – most surprising of all – a new heaven as well, filled not only with the multitude of stars at which people have gazed since mankind began, but equally full of radio-stars and X-ray stars which had been there all the time, unsuspected by any generation earlier than our own. We know all this only because an ordinary man like Glazer looked at an ordinary thing like a glass of beer and thought about what he saw in an *extraordinary* way.

Another man who did likewise was C. T. R. Wilson, a Scotsman who ended up as a Cambridge teacher (said by some to be one of the worst lecturers in the world). He actually came from an undistinguished village called Carlops near Edinburgh and, to the day of his death in 1960 at the age of ninety, spoke with a broad Scottish accent. At one time while still a student, he took a part-time job which involved climbing Ben Nevis to empty the rain gauge which was kept in the weather station on the mountain. Like all mountaineers he enjoyed the view of the landscape below him and then the climb through the clouds to the sunlit uplands above. But Wilson did more than enjoy the view and exult in the achievement of reaching the top. As he looked at the rays of the early morning sun striking the bank of cloud at his feet, he *thought about* what he was seeing. Wilson thought first about why clouds form the shapes they do. Why is it that up there in the blue there should be just one lovely little bouncing white cloud? What makes clouds form so definite a shape that the ancients long ago imagined cherubs with wings jumping about on them? It is not so long since people were half-convinced that when good folk die, they find themselves dressed in long white togas playing the harp while drifting about the firmament on a cloud. Be that as it may, it is undoubtedly strange that clouds have the diverse forms and shapes they do.

It will help us to think further about clouds if we first take a look at bubbles. Bubbles are spherical globules – balls, if you like – of gas within the body of a liquid. There is more to it than this, of course, and serious scientific research is in progress to this day on the more subtle aspects of bubbles, for example, on froth. Brewers investigate how to produce enough but not too much; people concerned with sewage, for instance, or certain chemical reactions try to find means to prevent froth from forming or to destroy it if it does form. If, however, for simplicity we accept bubbles as being spheres of gas in liquid, clouds can be thought of as tiny spheres of liquid (namely, water) in gas (that is to say, air). This, too, is an over-simplification but it nevertheless is a broad description of the facts.

Thinking about clouds, Wilson studied the way they are formed when the air is very damp and the temperature suddenly gets colder. There his talent for doing experiments came in useful. He was a slow worker. Indeed, he only did about three experiments in his life but each of these was so profound and incisive it was worth the output of a hundred diligent scientists of ordinary talent. He designed a simple cylindrical glass vessel with a well-fitting bottom which could be pulled down, thus giving it the general properties of a piston. If the vessel was filled with very moist air and the top closed, when the movable bottom was abruptly lowered, the air in the chamber was suddenly chilled by expansion. Cloud or fog did not necessarily occur. Yet when cosmic rays or man-made energetic particles were shot through the moist, chilled air, minute 'bubbles' of water could be seen to form on the ions left in the track of the flying particle, and the track of the particle could be seen as a line of fog.

Although Wilson's studies of clouds and his invention of what came to be called the Wilson 'cloud chamber' (the vessel with the piston-like bottom) undoubtedly added to our knowledge of clouds, the significance of the 'cloud chamber' was that, even before the time of the more complex 'bubble chamber', Wilson's comparatively simple apparatus provided one of the keys by which the previously secret characteristics of nuclear structure was unlocked. Nuclear physicists used the 'cloud chamber' to trace the tracks of particles of matter – protons and neutrons and, later on as discovery progressed, new kinds of energy-containing units undreamed of before. Their paths were traced in tiny streaks of fog. And, once again, as was demonstrated later with the 'bubble chamber', there were occasions when a moving particle might strike an atom fair and square and cause parts of its nucleus to fly off, each part also leaving a track of fog in its wake in the saturated vapour of the 'cloud chamber'. Conditions of humidity and temperature need to be exactly right for the trails of ions to form tracks of mist.

The white wake of a high-flying jet aircraft has a different origin. High in the sky where the jets travel, conditions are very cold. Jet aircraft derive the energy that keeps them flying from the combustion of fuel, petroleum. In chemical terms, petroleum is a hydrocarbon, that is, it is for the most part a combination of carbon and hydrogen. When the fuel is burned, the oxygen from the air combines with the carbon and the hydrogen, releasing energy and leaving a combination of carbon and oxygen, namely carbon dioxide gas and water. When a jet flies, streaming out carbon dioxide gas and water vapour behind it, the carbon dioxide is as invisible as the oxygen and nitrogen of which the air itself is composed. Things are different, however, for water vapour. Whereas on a warm day the water vapour that we breathe out in our breath remains invisible, if the weather is cold, it can be seen at once. The clouds of 'steam' we blow out of our mouth and nose on a frosty day are not steam at all but merely water vapour, in fact, tiny droplets of water. Where the jets fly, 30,000 feet or more up in the sky, the temperature is very cold indeed: the wake we see tracing the passage of the plane is composed of tiny ice crystals hardly bigger than dust.

The three events I have described, the tracks in a 'bubble chamber', those in a Wilson 'cloud chamber', and those made by a high-flying jet, illustrate the variety of problems to which anyone who claims to be a scientist must be alert. The 'bubble chamber' enables the track of invisible nuclear particles to be seen by making them form bubbles of gas in a liquid; the 'cloud chamber' allows one to see the tracks of the same sorts of particles because of the droplets of water they condense in gas; and the path of the aircraft is traced by tiny crystals of solid ice. But the example of the aircraft illustrates a very important point. The white tracks it leaves behind are not particularly significant, but it also produces an invisible track which, before its existence was known, took the lives of a large number of men and women.

What happens when an aeroplane rushes forward through the air is that the way the upper side of the wing is shaped causes the layer of air on the top of it to roll up into a spiral vortex – a kind of spinning corkscrew. It is the upward thrust of this twisting sheet of air that gives the lift to the aeroplane. As the plane flies along, the spiral vortex which is being twisted up (as it were) all along the wing, slides outwards (along each wing) and eventually spins off behind the trailing edge of each wing tip. This means that as the plane travels forward, it leaves behind it two 'sausages' of air, one spinning one way and the other spinning the other way. These are called helical vortices. This general effect has been known for some time. What is new, however, is the strength of these man-made 'tornadoes' (if we may call them such) and the length of time they last after the plane that made them has passed by. Their strength is directly related to the weight of the aircraft: it follows, therefore, that now that jumbo jets and the like may weigh up to 500 tons, the power and force of the vortices they leave behind can be very considerable.

By the time that about a hundred light aircraft (although some of them were quite big) had crashed and been destroyed because they had flown unsuspectingly through the invisible trails of a jumbo aircraft, experiments were carried out to study these vortices. This was done by causing big aeroplanes to fly through lines of coloured smoke discharged from various levels of a high tower. This made it possible actually to see the vortices left behind by the planes. And what was found was that three minutes after a plane had gone, that is, when it was seven miles away and often out of sight, the swirling sausage trails left behind from the tips of its two wings could still remain.

When I was a boy, I used to be told that the one thing better than presence of mind for a man in a railway accident was absence of body. This is also the best thing for light aircraft to avoid being capsized by the swirling, invisible 'sausages' of air. A number of experiments have been done to see whether it is possible – by modifying the tail fins or the shape of the wings of jumbo-sized aircraft, or even by building special bumps or protuberances on the fuselage – to stop the swirl of air from forming or at least to make it less lethal. But while some progress has been made – at some cost to speed and efficiency of operation – the best solution is still for another aircraft not to be there at all. It is worth the slight risk of a few irate passengers complaining about the late departure of a flight.

It is easy to forget that we swim through our life immersed in a river of air. For example, every schoolchild is taught that everything falls at the *same* speed and that this chap, Galileo, dropped a ping-pong ball and a golf ball off the Leaning Tower of Pisa and found that they both reached the ground together even though the ping-pong ball was lighter than the golf ball. As it happens, Galileo did not do exactly this. But he did say that a light object and a heavy object would both fall at the same speed if they were not hampered by the air in which they were falling.

There is some similarity between the problem of the golf ball and the ping-pong ball and the old chestnut which runs, 'Which is heavier, a ton of lead or a ton of feathers?' Although the answer is that they both weigh the same (a ton is a ton, no matter what it is made up of), it is still a good idea, if one or other is to be dropped on your head, to choose the feathers rather than the lead, particularly if you are allowed to arrange for them to be dropped loose rather than tied up in a sack.

Scientists have been interested in the question of falling objects for quite a long time. In fact what follows was covered in some detail by Galileo – described on the title page of his book published in 1610 as 'a gentleman of Florence, Professor of Mathematics in the University of Padua'. He it was who pointed out, that although a ball made up of a pound's weight of gold would fall at speed and land with a thud killing anyone who happened to be standing down below, if this same pound of gold were beaten out into very thin leaves, as it is when it is to be used for gilding, instead of falling like a stone, this same weight of metal would flutter delicately down like so many feathers.

Or consider the case of the load of canaries in a lorry with soft tyres. The driver and his mate set out to drive from London to Edinburgh. In the noise and bustle of London all goes well. Out in the country, however, the lorry begins to labour. The driver stops, his mate gets out, goes round to the back and bangs on it with a plank. This frightens the canaries inside, they all fly up, the weight on the tyres is reduced and the lorry drives on as merrily as ever – until the birds settle on the bottom again.

One more experiment before we come to the answer. Stand on the top of a pair of steps with a beach ball in one hand and a gas-balloon the same size in the other hand. When the two are let go, the beach ball falls *down* while the balloon falls *up* (or at least it rises up into the sky).

The truth is that in the universe at large, everything does fall at the same speed when things fall at all. To put the matter more precisely, everything is subject to the laws of gravity and, when there is nothing to interfere with its doing so (and this is the point), heavy things and light ones get up speed with the same degree of acceleration and take the same time to get down from the top of the cliff to the bottom. In real life, however, and on the real earth there *is* something to interfere with things falling. This is the air. If the lorry and all the canaries in it were on the moon (where there is no air and the rocks, stones and spaceships are all sitting in a vacuum), the canaries could not fly and would all be on the bottom, so that the full weight (such as it is on the moon) of lorry and canaries would rest on the tyres. The problem would therefore not arise. Similarly, if down here on earth, a vacuum were pumped out in a great pipe standing on end to the height of the Leaning Tower of Pisa, the ping-pong ball and the golf ball would fall at the same speed inside it. Even more remarkable would be the sight of the beach ball and the gay-balloon hurtling down the evacuated pipe side by side and crashing to the bottom. Even the one-pound ball of gold and the sheets of delicate gold-leaf would drop like stones down the pipe together.

Perhaps it is just as well that this is not a 'perfect' world (perfect, that is, for the people who involve themselves with gravity) and that there is air in it to buoy us up. It is true that big jet aircraft would not leave a vortex behind them as a danger to following planes – but then there would not be any aircraft if there was not any air. Badminton would not be the same either since the shuttlecock would travel like a bullet and fall like a plummet. On earth, a mouse dropped over a cliff runs away when it gets to the bottom, a dog lies howling with a broken leg, and a cow splashes. Without air even the mouse would come to grief.

The 'bubble chamber' invented by Dr Glazer and the 'cloud chamber' invented by C. T. R. Wilson made use of bubbles and cloud to show where the particles of matter were going and what they were doing. The sausage-shaped vortices left behind by passing jumbo-jets cannot be seen, yet they represent substantial forces in the invisible air. Last of all, the atmosphere holds up our aeroplanes and slows down the mouse as he drops so that when he lands he will not be hurt. And yet, invisible but strong and unbreakable, the laws of gravity stretch through the air, just as they stretch through space.

And all this comes out of some good thoughts about the bubbles in beer.

6 Stay-put stuff

For quite a long time – that is, for hundreds of years – people have asked themselves what kind of stuff is stuff. What really *is* wood and stone, paper and bricks, a leg of mutton and a glass jar? It's not enough to say that wood is made of carbon, hydrogen and oxygen, that stones are calcium and magnesium, that glass is silicon. This is useful information, no doubt, but it leaves out the basic question. Everything we see and touch possesses certain recognizable characteristics. One might almost say that things, solid bodies, if you like, or – more important still – human bodies as well possess a spirit or soul that we recognize as being the very essence of every kind of stuff on earth. What am I talking about? I'll tell you.

There you are, sitting in a bus. On your knee is a brief-case. At your feet lying on the floor is an empty beer bottle. In your pocket are five 10p pieces. The bus is merrily travelling along at an even pace. Suddenly a dog runs out into the road and the bus driver slams on the brakes and stops. While this is happening, what are you doing? As the driver abruptly pulls up, you jerk forward and, for all I know, bump your face on the back of the seat in front. And while you are doing this, the brief-case has shot forward and fallen on the floor and the beer bottle has rolled all the way along the bus until it bumps into the partition separating off the driver at the front. Why do these three quite different things, the brief-case, the bottle and you yourself, behave the same, and what is it that makes them behave as they do?

The answer is that all things, that is to say, all the material stuffs on earth (and, as a matter of fact, most of the materials that make up the sun and the stars as well) possess in their substance an odd but universal influence (a mysterious central core) which scientists call *inertia*. Even the 10p pieces in your pocket have this same inertia in them and go on forcing themselves forward after the bus has stopped.

Young Isaac Newton, that serious, thinking boy who grew up into a serious, thinking and rather disagreeable man who was one of the greatest geniuses England has ever produced, was the individual who first spotted what kind of quality inertia is. It is related to the amount of whatever one is concerned with, the mass of stuff present, as scientists would call it. The main feature of inertia, be there much or little of it, is that if a mass of stuff is still, the inertia that is part of whatever stuff it is has a tendency to keep it still. On the other hand, if a mass of stuff is moving along, as you, your brief-case, your beer bottle and your money all were in the bus, the inertia in all these things has an equal tendency to keep them going whether the bus stops or not.

There are all sorts of ways of showing what inertia is, how much there is of it and what it does. For example, it is a matter of common knowledge that it takes much more effort to *start* a barrow, or a motor car, or a railway truck going than it does to *keep* it going. The reason is that since inertia exerts a force keeping things still that are already still, this inertia has to be overcome before the truck or the car can be got going. Of course, when it *is* going, its inertia is on your side because, as Isaac Newton also said, when anything is on the move, its inertia then exerts a force keeping it on the move. This can easily be shown with a toy motor car. Tie a piece of elastic to the front and pull it steadily. The elastic stretches longer and longer while the toy car stops where it is. At last, when the force of the inertia keeping it still is overcome by the pull of the elastic, the car starts. But once it *has* started, very much less force has to be applied (as can be seen by the way the elastic is less stretched) to keep it steadily on the move.

Inertia is not only present in – or, should I say, belongs to and is part of – every thing and stuff there is, but it works just the same whether things are moved up and down or backwards and forwards. Should a man be pushed off a cliff, he hesitates in mid-air for a moment before he starts to fall down to break his neck at the bottom. The force of gravity pulls him *down* in just the same way that the piece of elastic pulls the toy car *along*, but before it can get him started on his downward drop, it has to overcome the inertia in his body. If you are quick enough, therefore, you can, after pushing the man over the cliff, change your mind and pull him back again.

Inertia to the pull of gravity can be neatly demonstrated by hanging two weights on a bar by pieces of thread and extending the threads beyond their attachments to the weights so that the tail ends hang below them. If one takes hold of one of these ends and steadily exerts an increasing pull downwards, the point will come when the part of the thread *above* the weight will snap and the weight fall down. This is because the weight of the weight combined with the additional pull which has been continuously applied eventually becomes too great for the thread to resist. On the other hand, if one takes the thread dangling below the other weight and gives it a *sharp* pull downwards, it will snap *below* the weight, which will remain hanging from the bar. This happens because the sudden tug is resisted by the inertia of the weight, which has a tendency to stay where it is that is sufficiently strong to break the thread.

For those who would like to illustrate inertia in a dramatic way, demonstrating its force and its widespread distribution in everything, an experiment with a broomstick and two wineglasses is to be commended. The wineglasses should be stood on the seats of two wooden kitchen chairs six feet apart. The broomstick must then be balanced as a bridge from the nearside rim of one glass to the nearside rim of the other. It is, perhaps, advisable to insert a needle into each end of the broomstick and rest these on the wineglass rims to avoid damage from the roughness of the wood. The demonstration can then be carried out. Take a heavy poker or a wooden club and bring it straight down with full force on to the middle of the broomstick. The stick will be broken to pieces but the wineglasses, each on its own chair, will still stand intact.

The explanation of this remarkable demonstration is that while the middle section of the broomstick starts downwards with speed under the abrupt and overwhelming force of the descending poker, the inertia of the two ends resisting any proposition to move them does so with a force sufficient to break the rather rigid broomstick. Hence, since it is this resisting, 'stay putting' force which breaks the stick, the wineglasses, which are at no time called upon to stand up against any mechanical shock at all, remain unscathed.

A variant of this experiment which perhaps illustrates the power of inertia even more vividly was popular a hundred years ago or more. The learned professor doing the experiment called out from his class two of the youngest of his pupils, stood them about six feet apart and gave them each a cut-throat razor to hold. He then hung a paper ring over the blade of each razor so that it rested on the razor's edge which was directed upwards. Finally each end of the broomstick was passed through the paper ring so that it was supported in a horizontal plane. When with all his strength the professor struck the centre of the broomstick with his club, it was broken to pieces while the paper rings were undamaged and not even cut by the razors.

Just as in this experiment it is the property of inertia residing in the ends of the broomstick which allows wineglasses to stand unbroken in the crash of splintering wood, so an alternative experiment can be done to demonstrate the potency of the inertia in domestic crockery. To do this, cover a smooth polished table with a smooth flexible tablecloth. The table may then be laid with plates and cups, teapots, sugar basins and serving dishes. Since the amount of inertia in an object is related to its weight, the ease and safety with which the demonstration is carried out are enhanced the heavier the items are with which the table is set out. The scientific demonstration is for the bold and self-confident operator to grasp the tablecloth firmly by the two corners of its front edge and, with a vigorous horizontal jerk, pull it off the table. The inertia which constitutes an essential quality of each plate, cup, dish and pot will influence them to stay where they are – unbroken on the bare table.

Taken all in all, inertia is peculiar stuff. That is exists is incontrovertible as the simple experiments I have described make so abundantly clear. All we can say is that it is a part, an invariable component of the stuff that we and everything else are made of. For those with a taste for mathematics, it is possible to go into a great deal more detail. For example, the amount of inertia there is in any particular thing or system can be worked out. The more there is of a thing – or, as the scientists put it, the greater its mass – and the faster it goes, the greater becomes the inertia. There is some truth in the common saying of people descending in a high-speed lift from the twentieth storey of a tower block that they have left their stomach behind. The inertia of a stomach heavily laden with lunch would, if it could, keep it still on the twentieth floor. And the disagreeable sensation at the end of the ride is again caused by the inertia of the traveller's inner organs which – as is the nature of inertia – exert a force aiming to keep them travelling on in the direction of descent even after the lift has stopped.

And this brings us to the next point. Isaac Newton pointed out that the inertia in a freely moving body tends to make it go on moving *in a straight line.* If it is compelled to follow a curved path, its inertia exerts a force against whatever it is that is compelling it to curve and it will, therefore, constantly try to 'fly off' at a tangent. This tangent will be the straight line that the inertia of every moving body wants to follow.

Now consider a child's idea of Australia and the Australians. There can hardly be a boy or girl in the entire continent of Europe who has not at one time or another wondered how it is that Australians, walking about with the soles of their feet pointing towards us, manage to stick on and prevent themselves from falling off the earth. This is a good question to ask just as it is a difficult one to answer. It is all very well to talk about gravity just as it is useful to work out what can be called the 'laws of gravity' describing the force with which bodies of different sizes pull each other together. Yet for ages, indeed ever since it was first discovered that the earth is not flat but round, while the mathematics of gravity were studied and worked out in considerable detail, no one had any clear or plausible ideas about what gravity was. The first thing to notice about gravity is that when something is dropped, whether it is a stone or an apple, it falls down and, as it falls, it gets up speed. that is to say the rate at which it drops accelerates. The next point, which is a curious one and rather difficult to demonstrate – although it is true nevertheless – is that no matter what the weight of an object may be, one thing falls at the same rate as any other. That this is so has been abundantly proved yet, should one drop a golf ball and a ping-pong ball from the top of a pair of steps, the heavier golf ball will get up speed faster than the lighter ping-pong ball. But this is nothing to do with gravity; as explained in the last chapter, it is caused by the drag exerted by the resistance of the air. Although the equal acceleration of all kinds of things regardless of their weight can be demonstrated by dropping them in a vacuum, this is rather difficult to organize. An easier way is to slow up the rate of fall and thus minimize the effect of the air resistance by allowing the two balls of different weight to roll down sloping runways set up at exactly similar angles. This has the added advantage of making it easier to observe and time what happens.

The fact that big things, little things, solid massive stuff like lumps of lead and light porous material like corks all get up speed – that is, accelerate – at the same rate when they fall raises a number of odd, peculiar and interesting points about the nature of things and of the inertia built into their substance. For example, one can do experiments in the cabin of a jumbo-jet flying over the North Pole or in a first-class sleeping compartment of a crack railway train speeding smoothly along its welded rails. One can drop golf balls and ping-pong balls, roll them down slopes or bounce them on the floor and obtain precisely the same results as if one were carrying out the trials in the peace and calm of a laboratory standing, firm as rock, in an ancient city of dreaming spires. Only if the pilot of the aircraft drops his flaps and shuts down the engines – that is to say, causes the plane to decelerate – or the engine driver suddenly gets up speed does the scientist in the plane or in his compartment in the train really know (particularly if he keeps the blinds drawn) whether he is moving or not. This, then, brings us to the next experiment which is one invented by Albert Einstein.

The man doing the experiments is now in a space capsule speeding along at a steady pace where there is no gravity. In one hand he holds a wooden ball and in the other hand a lead one. The whole system, the capsule, the man and the balls are speeding steadily along. As his first experiment, if you can call it such, he lets go of both balls. There being no gravity up in space and the inertia of the balls being what it is, they go sailing along just as they did before. Of course, to the man in the capsule who has just let go of them they appear to stay still in 'mid-air'. He then does the next experiment.

He goes to his auxiliary rocket motors and accelerates the speed of the rocket. What then seems to him to be happening is that the two balls immediately start to fall down in the capsule until they strike the floor. If he adjusts the rocket motors so that the acceleration of the capsule becomes equal to that at which things fall on earth, he can even confuse himself with the notion that he has never left the ground. But although to the man in the cabin of the space rocket the balls that had appeared to be stationary suddenly fell down, to an observer watching the whole thing from outside the rocket what happened was that the balls, steadily travelling through space in a straight line kept on going at the same speed by virtue of the inertia in them while it was the floor of the capsule that suddenly started to go faster, overtook and bumped into them. Furthermore, the floor of the capsule will bump into both the wooden ball and the lead ball at the same moment.

Looking at things from the outside, we can see that the astronaut now standing firmly on the floor of his capsule feels that he is still only because the continuous acceleration of the rocket is continuously pressing the floor up against his feet and he thinks the balls are falling because they are being overtaken as their inertia keeps them steadily on course. People who talk about astronauts and the conditions in their orbiting space laboratories often refer to the 'artificial gravity' produced, either by getting up speed or by making the vessel spin. But is this gravity so artificial after all or is it merely this pervasive, peculiar quality, inertia, that is an essential feature of everything?

But now we come to the big question. If a man *thinks* that he is standing still on the floor of his capsule because, when it continuously accelerates, the 'gravity' he feèls (due to the push of the rocket motor against his built-in inertia) is the same as the gravity he experienced on earth, how do we know whether or not the gravity we feel on earth is not due to the same cause? In other words, the laboratory in the basement of the solid university building – as firm as a rock – may in fact (if there was an outside observer to see what was going on) be speeding along, accelerating as it goes, just like the man in the capsule. That is to say, the firm rock, the laboratory, the whole world in fact, may be rushing through open space and not be firm at all. When things are looked at in this way, the only *real* thing about anything – the world, lead balls, people and all – is the inertia it possesses.

Everyone at some time will have swung a bucket of water over their head. If the bucket is kept swinging and not bumped against an overhanging branch of a tree or knocked against one's own leg, the water will stay in it even when it is upside down. The reason for this is that its inertia tells the water to continue to go rushing on through space at a steady speed in a straight line. On top of this, having been started in one direction by the swinging bucket, the water's inertia tells it to go on in that direction, but because the bucket is made to follow, not a straight path but a circular one, the water has to get up speed – to accelerate, that is – to get round the corner. This gives it 'artificial gravity', if one likes to call it such, to enable it to stick to the bottom of the bucket even when the bucket is upside down.

I have already tried to explain that, provided we accept that the ghostlike inertia that everything possesses is real, whatever else may be an illusion in this world, then it does not matter that the earth is not stationary at all. But even so, if it is moving, it may not be travelling like a rocket on a straight journey from A to B. We have already seen that things behave exactly the same whether one is standing still in a basement, standing in the

'Everyone at some time will have swung a bucket of water over their head.'

first-class dining saloon of a transatlantic liner, or in a plane travelling at 700 m.p.h. over the ocean. One can drop a tennis ball or bounce it in any of these places and it drops and bounces just like it does when we know we are standing still on firm ground. Suppose, however, one were in a closed car going round and round one of those enclosed circular towers one sees at fairgrounds called the 'wall of death', would things be the same then? Or, to be less extreme, suppose one was expected to live for a few months in a sitting-room built on to the platform of a continuously revolving carousel or merry-go-round, what then?

If a man is standing with a blindfold over his eyes on the floor of a merry-go-round anywhere other than at the exact centre and supposing that the merry-go-round is revolving steadily, he will feel a force tending to move him towards the outermost edge of the platform. A ball left to roll will likewise roll towards the outer edge of the merry-go-round. But if the blindfold man did not know he was on a revolving roundabout he could ascribe his sensations and the motion of the ball on the floor equally well to his standing still on a *sloping* floor under the ordinary force of gravity. What is happening is that the man's inertia and the inertia in the ball, possessing as they do the built-in tendency to continue going in a straight direction are exhibiting that tendency in resisting the circular motion of the merry-go-round. And by doing so, they produce an imitation of the force of gravity. Or, to put this the other way round, the force of gravity, by which the stars move, the earth revolves in its seasons round the sun, and apples fall off trees, can be explained in terms of this single, strange, universal quality of stuff, inertia, provided we come to terms with the fact that we and the little universe we know about are not still at all but rushing through space and that space itself may not be a straight tunnel leading from A to B but may be curved.

Isaac Newton worked out his early ideas about the universe by watching an apple drop off a tree in his uncle's garden when he was evacuated from London to avoid the plague. There is no reason why we should not follow his example by coming to an understanding, not of *what* happens but *how* it happens, by pulling the tablecloth off the teatable without smashing the crockery or shattering a broomstick without breaking the two wineglasses supporting it.

If all this – simple as it may be even though it can lead to an insight into the inner nature of things – is one aspect of the heart of matter, one can come at the answer in another direction which also has a rather unexpected starting-point.

Amber is a pleasant-looking, yellow, translucent material not very commonly seen nowadays. It is used for making the mouth-pieces of old-fashioned pipes. It also makes rather elegant beads and brooches or it can be carved into little figurines to put on one's mantelpiece. Anyone likely to be involved in a quiz game may like to know that it is a fossil resin and that most of the amber one sees has been dug up from the 'blue earth' in what was once Samland, now Poland. One can, however, find pieces of it at various places along the east coast of England, on the coasts of Holland, Denmark and Sweden and in south Finland as well as, here and there, in other parts of the world. The reason why I am bringing the matter up becomes apparent when I mention that back in 600 B.C. the intelligent and civilized Greeks had a word for it: and the word they used for amber was *electron*. This word, which is used today with a special meaning by engineers, scientists and by ordinary people as the proper basis of electricity and electronics, points just as deeply as *inertia* to what we are getting to know about the nature of things. Yet what the Greeks saw in 600 B.C. and what we see today is as simple (or as mysterious) as it ever was.

What the Greeks saw was that if you rubbed a rod of amber with a dry cloth it would attract little bits of paper or leaves and, on a dry day, would give up this power with a crackle and a spark. Though amber has perhaps the most potent ability to do these things when rubbed, other materials do the same. Therefore, what we hear when on a dry night we pull a woollen jersey over our heads is our hair crackling. A nylon vest will do the same with even greater force and it is commonplace to produce sparks and crackling by combing one's hair with a nylon comb. What we are seeing is static electricity trying to get out. 'Where does it come from?' you may ask. The answer is this.

Not only are sticks of amber, silk, cloths, woollen jerseys and nylon undervests made of electricity, but so is everything else as well. Normally the electricity things are made of stays quiet because the positive electricity and the negative electricity are present in equivalent amounts and consequently cancel each other out. But when the amber rod is vigorously rubbed with a woollen cloth, the cloth rubs some of the electrons off the rod and (electrons being negatively charged) the amber becomes strongly charged with positive electricity. Such a change represents a state of tension. This state of tension struggles to release itself by grabbing electrons from wherever it may, from the hand of anyone who touches it. Bring it near some light object such as a scrap of paper and so avidly does it attract electrons that it draws up the pieces of paper, not by the hair, but by the electrons in them. The odd part of the business, however, is that electrons were part of the amber in the first place. Amber being a bad conductor, when the electrons were rubbed off, the positive charge left behind remained in it.

Run for a bus on a cold dry day and the friction of your arms and legs going up and down rubbing the material of your clothes will knock off electrons and generate a charge so that when your hand grasps the rail of the bus, there is a smart click, a spark (if you can see it) and you feel a nasty shock. People in Great Britain are less bothered by this sort of thing than people in, say, Canada where the air (particularly in the winter) may be very dry. Because it is usually damp in Great Britain, any charge tends to be dissipated by way of the moisture in the air.

As I have said, people have been interested since ancient times in electricity and the way it can become accumulated in certain materials which are, as we say, bad conductors. For at least 200 years ways have been devised by which quite large amounts of electricity – 'static' electricity, as it is called, because it stays still, at least until it finds some means to escape – can be collected. The simplest kind of device uses a disc rather than a rod. By mounting a disc of glass or resin – a record or a nylon disc would do equally well – on a spindle so that it can be revolved with a pad rubbing against it, a more effective way of applying friction is provided than merely by rubbing a rod. The charge of electricity can then be encouraged to jump off the disc by fixing a row of metal points close to it. A man standing on some sort of insulating material such as glass, plastic or even a barrel of pitch and holding a metal knob connected to the row of points picking up the charge of the revolving disc will himself be charged with electricity and will receive a considerable shock if he then touches some object – a metal pipe, for example, by which the charge can escape to earth. A pleasant variation of the same experiment is to charge a lady with electricity from a frictional machine, as it is called, and then encourage a gentleman whose feet are on the ground to kiss her. Both will receive a shock as the charge leaves her and passes through him to the ground.

The generation of electricity by doing work is now well understood. We think nothing of touching a switch to make wires white-hot in the glass bulbs that give us light. We likewise make columns of gas incandescent in neon lamps, run the trains and work the factories. There is a mass of learning known as electrochemistry by which analysis is done and materials of all sorts are made. And out of all this comes the knowledge that the solid earth and all that's in it, the glass, the iron, the wood, the stone, are not exactly the massy things we think they are. When we magnify what we see first with a microscope, then with an electron microscope, then with X-ray crystallography and then with tools even more sophisticated, we understand – it would not be quite accurate to say we see – that what appears to be solid material is not so at all: there is as much empty space in a rod of iron, not to mention a beam of wood (which is mainly air anyway), as there is in the heavens of a starlit night. And the particles in the tiny bits of stuff that revolve in their minuscule orbits of empty space are themselves only electromagnetic particles.

But if this sounds altogether too high-flown and schoolmasterish, one can take comfort in the understanding we can get from what we can so easily see and feel. When we come down to it, our own observation shows us that at the heart of the matter – that is every kind of stuff on earth – is *inertia*, this curious tendency to stay still or to go on going on. It is, after all, real enough to keep the crockery on the table when the cloth is pulled off – or to illuminate the laws of gravity. And the second thing in which we can believe is that, if you rub wool or silk or even leather against glass, nylon or – as has been known for the best part of 3,000 years – amber, you will rub off some of the electrons, the outer layer or skin, of which the glass, nylon, leather, or what you will is made. What happens is now attributed to 'quantum mechanics' and is complicated enough. But it shows that, besides the *inertia* which we cannot see but know from its ghostlike pulling and pushing that it is there, the solid substance of everything is made of *electromagnetism* which also we cannot see but which, like inertia, can also make itself felt.

7 Bathtime in Nanuki

Wherever one may be in the Northern Hemisphere, from Norway to Algiers, from Alaska to Mexico, or from Novosibirsk to Tokyo, if, after having taken a bath one carefully removes the plug, the water as it starts to run away will soon form a vortex, spinning round as it goes. In all these places, assuming that the baths have been installed with proper care, more often than not the spin of the water as it gurgles away will be left-handed. That is to say, the twist of the vortex formed over the plughole is anticlockwise.

But if you are in Buenos Aires or Lima; in Perth, Western Australia; in Auckland, New Zealand, or in Dunedin the bathwater left to itself will mainly run away spinning in a clockwise direction. Only in a few centres of population where the plughole of the bath is directly on the Equator, such as Nanuki in Kenya, will the bathwater run away twisting as often one way as the other – or often enough not forming a vortex at all when the plug is pulled out.

To ask whether it matters which way the water runs is neither here nor there. To a scientist the question behind every phenomenon, every quirk of nature – the glittering piece of rock; the bat flying in the dark without bumping into things; the oddities of starlings' behaviour; stars twinkling; whales apparently singing to each other; or the origin of all that oil under the sands of the Holy City of Mecca – all these questions are equally important or unimportant. But most important and fascinating, delightful, and often baffling and frustrating as well, is finding the answer why.

The bathplug business is particularly interesting for two main reasons. First of all, it is quite difficult to find out the facts: does the water always twist one way and, if it does, which way does it twist! And then, why does it spin the way it does? If everyone reading this sentence put the book down, filled a bath with water, pulled the plug out, watched what happened and described what he or she saw, a confusion of results would be ventilated and a clamour of argument raised. Nevertheless, just as the chances are, for every 10,000, 100,000 or 1,000,000 babies born there will always be more boys than girls, so (in the Northern Hemisphere) there would always be found to be more left-hand-revolving bathwater. There is in fact a tendency, an influence, if you like – that is to say a natural force operating in all the bathrooms in half the earth – that twists the water leftwards as it runs away. Truly it is a lovely thing that there have been (and still are) true and dedicated scientists who – rather than devoting themselves to the formulations of nondrip paint or high explosives have applied their talents to establish the fact that this influence really exists.

The classical modern research on bathwater was carried out in 1962 by Professor Ascher H. Shapiro, Head of the Department of Mechanical Engineering at the Massachusetts Institute of Technology.[1] Professor Shapiro did not use an ordinary bath in an ordinary bathroom. Nor was he satisfied to use ordinary bathwater. Had he done so, all sorts of interfering influences could have disturbed the results of his experiments and smothered whatever natural influences existed. For example, any swirling of the water that might have been going on before the plug

1 Shapiro, A. H., *Nature*, 196, 1962, 1080.

'. . . whales apparently singing to each other . . .'

was pulled out could have upset what happened afterwards. Even if the water was still, it could be unevenly heated, with a hot part one end of the bath and a cold part the other. Or a draught under the door might be blowing on the water and causing it, no matter how little, to swirl.

To avoid all these disturbing factors, Professor Shapiro used an enormous circular saucer, six feet across but only six inches deep, with a hole three-eighths of an inch in diameter exactly in the centre. He took great pains to fill the vessel in a uniform way: to protect the surface of the water from draughts he did not even allow any direct light to shine on it lest it be warmed in patches and, most important of all, he allowed the water to settle quietly for twenty-four hours before carrying out any of his experiments. When all these precautions (and some others which I shall describe later) were taken, Professor Shapiro found that the water always circulated in an anticlockwise direction as it drained away. Professor Shapiro's laboratory was in the Northern Hemisphere – at Watertown, Massachusetts.

The publication of Shapiro's results in the scientific literature did not finish the matter. Two years later over in Britain, Dr A. M. Binnie, F.R.S.,[1] thought it worth while to stop what he was doing in the Engineering Laboratory at Cambridge to find out whether he could repeat Shapiro's results using a tank five feet in diameter and three feet deep. To start with everything went wrong: conflicting results occurred when he only allowed the water to settle for three hours; a special baffle he had fixed on the outlet pipe did more harm than good; the water from the Cambridge town main he was using formed scum and bubbled after standing for the necessary twenty-four hours. Dr Binnie gave up his first series of experiments when he found that the cellar where he had his tank was so draughty that in cold weather the water began to revolve by itself.

In the second series Dr Binnie took extraordinary pains to make sure that the way the water was kept in the tank had no effect on what happened when it was run out. To keep out dust and draughts he kept the top of the tank covered with polythene sheets the whole time. When he looked to see what

1 Binnie, A. M., *I. Mech. Eng. Sci.*, vi, 1964, 256.

was happening he peeped in through a narrow slit between two of the covering sheets. He did not allow electric light to be used in the laboratory for fear of its heating effect and relied on a north-facing skylight. He let the water settle quietly for a minimum of twenty-one hours, and for some of his experiments even allowed three days before he pulled the plug out. After all this his experiments were a complete success. The same thing happened every time, entirely supporting Professor Shapiro's experiments: during each of fifteen trials the water ran away spinning in an anticlockwise direction.

The next stage in the investigation was predictable: the scientific community was roused. Within a year a report appeared describing what happens to Australian bathwater at the University of Sydney.[1] Benefiting by the experience of Shapiro and Binnie, the Australian scientists used a circular tank six feet in diameter and nine inches deep with a central outlet hole. Their vessel was of plywood, not metal, to reduce the chance of temperature variations in different parts of the tank setting up eddy currents. The drainpipe was not fixed flush with the bottom but projected one inch up; this was to avoid any possibility that the bottom might exert some braking effect capable of influencing the circulation of the water, one way or the other, when it was allowed to run away. The tank with all its fittings was set up in a small, cement-walled room in the basement of the university. This room had no windows and was ventilated from an overhead grille. The temperature of the room and of the water running into the tank was controlled by a thermostat so that it never varied by as much as one degree from 20°C throughout the entire series of tests. The tank was filled through a hose to a depth of six inches above the top of the outlet pipe, and since they expected the circulation to be clockwise, being in the Southern Hemisphere, they directed the hose so as to leave the water swirling anticlockwise after the tank had been filled.

When everything was ready and the tank filled, Trefethen, Bilger, Fink, Luxton and Tanner set to work. They blocked off the ventilation louvre in the ceiling with plywood and covered

1 Trefethen, L. M., Bilger, R. W., Fink, P. T., Luxton, R.E., and Tanner, R. J, *Nature*, 207, 1965, 1084.

the plywood with a plastic sheet; they covered the tank itself with two plywood sheets, only leaving a small slit between for observation; they went away and allowed the water to settle for eighteen hours; when they came back, they slipped in through the door as quickly as they could and then kept the door closed while the experiments were being done. And what did they find? The water always ran away down the plughole twisting in a clockwise direction, that is to say, with a right-hand spin when one looks down on it from above.

It can with some confidence be asserted after the diligent work of these seven devoted scientists that bathwater runs away anticlockwise in the Northern Hemisphere and clockwise in the Southern Hemisphere. In 1974 Yorkshire Television in a programme in which I was involved organized a single confirmatory experiment – not done with all the rigours of science as in America, England and Australia, but an experiment nevertheless. There is in the small town of Nanuki in Kenya an unpretentious hotel which has one particularly notable feature: the plughole of its bathroom is situated exactly on the Equator. In the experiment, a bath in this hotel was filled and confetti sprinkled on the surface of the water. When the plug was pulled out, the confetti circulated neither one way nor the other but went straight down the drain from all sides.

What, it may be asked, was the use of all this effort in America, Australia and England, not to mention the single observation in Kenya? Were these scientific inquirers wasting their time? To a true scientist, the answer is no. These quite extensive researches, even though they were devoted to so apparently trivial a matter as bathwater running away, illuminate a general principle: whether the water is in the privacy of a bathroom north or south of the Equator or bang on it, it is whirling round the earth's axis.

When a man stands on the Equator the solid ground under his feet is rushing towards the east at a speed of approximately 1,000 miles an hour. Farther north, where the earth is not so fat, a man standing on the surface is moving slower, as he has less far to travel in each time of revolution. For this reason, the surface of England only travels eastward at about 700 miles an hour. Imagine an enormous bath with one end at the Equator and the other at Greenwich: the water at the southernmost end is moving faster than the water at the northernmost end. It would follow, therefore, that when the water got a chance to do so, as it would when the plug was pulled out, it would have a tendency to revolve counter-clockwise. Although a real bath is very much smaller than the bath we have been imagining, the same principle will apply to the water in it. That is to say, that part of the bathwater lying to the south will be flying through space faster than that across the bath on its northern side. The difference in speed will not be very much (hence the difficulty experienced by all those scientific gentlemen in America, England and Australia in finding out definitely what was happening), nevertheless it is there. And it is not too fanciful to say that the waste pipe of a bath is a useful device for demonstrating the rotation of the earth.

It takes rather special talents to feel the earth spinning underfoot but if one is bright enough it is not difficult to set up an experiment to demonstrate it. This was in fact neatly done by Jean Foucault in Paris in 1850. He suspended a heavy weight on a long cord from a hook fixed in a high ceiling so that it hung exactly over the centre of a graduated circle similar to a compass dial painted on the floor. If such a weight is set swinging across the centre of the circle, it can be seen at the beginning to pass backwards and forwards along the middle line dividing the circle in half. If the circle is graduated into 360 subdivisions all the way round, the weight will swing from 0° to 180° and back, over and over again. Soon, however, if the weight hangs steadily from a firm support and is unaffected by draughts, it will be seen to be swinging not from 0° to 180° but from 1° to 181°, then from 2° to 182° and next from 3° to 183°. What is happening is that as the weight swings freely backwards and forwards, the world is revolving under it. Because the weight is hanging freely moving to and fro under its own momentum, there is no reason for it to twist round with the earth. It could even be argued that the weight is swinging steadily backwards and forwards and that it is the ground underneath it with the circle painted on it that is turning round.

It is really rather satisfactory: we now know that in a quiet house by hanging a weight on a string over the second-floor banisters anyone can actually see the world go round.

The arrangement of a swinging weight is now called a Foucault pendulum in honour of its inventor – he deserves the credit.

Once a man or woman begins to take a serious interest in the way the universe works – that is to say, in science – there is no telling what may turn up. And what Foucault turned up with was a remarkable observation with a gyroscope. At the time, he believed that he had invented the gyroscope but although he was about forty years too late, the experiments he did with it were not only remarkable for what they showed but also produced results which are still demonstrating their usefulness today.

It was a German scientist called G. C. Bohnenberger who made the first gyroscope in 1810. It was not quite the same as a modern model: instead of a heavy wheel Bohnenberger used a squashed metal ball, shaped somewhat like a tangerine, and fitted it inside a metal ring so that it could be spun. The ring with the metal ball mounted in it was itself mounted on pivots inside a second metal ring arranged at right angles to the first. This second ring was in its turn balanced on pivots inside a third. This meant that when the ball was set spinning, it was free to tip in any direction. A number of people played with this glorified top for the best part of twenty years. It behaves in a fascinating way, as children (quite apart from scientists) soon discovered. In 1836, however, Andrew Lang got the notion that a gyroscope could be something more than a sophisticated puzzle or a scientific toy and (like bathwater running down a drain) might be used to demonstrate the rotation of the earth on its axis. This is what he wrote:

'While using Troughton's top [this was a modification of Bohnenberger's design] an idea occurred to me that a similar principle might be applied to the exhibition of the rotation of the earth. Conceive a large flat wheel, poised on several axes all passing directly through its centre of gravity, and whose axis of motion is coincident with its principal axis of permanent rotation, to be put in very rapid motion. The direction of its axis would then remain unchanged. But the direction of all surrounding objects varying, on account of the motion of the earth, it would result that the axis of the revolving wheel would appear to move slowly.'

But though Andrew Lang had the bright idea, he did not do anything about it. This was left to Foucault. Actually, it is doubtful whether he knew anything about what Lang was saying in Scotland or even what Bohnenberger had done in Germany. It is more likely that he invented the gyroscope afresh for himself and he was certainly responsible for giving it its name. He carried out his experiment in 1852 and when it was demonstrated at the Liverpool meeting of the British Association for the Advancement of Science in the same year, it made an immense sensation. People were astonished at the notion that a spinning top (which is basically what a gyroscope is) obeys the same laws in space that control the stars and goes on doing so independently of the revolving earth on which it happens to be placed.

The first gyroscope experiments were difficult to do because no matter how fast a heavy gyroscope wheel is set spinning, sooner or later it comes to a stop. This may happen before the earth has had the chance to revolve very far. It is therefore not surprising that the use of a spinning wheel to act as a ship's compass was not achieved until after about 1906 when a way to keep the wheel spinning almost indefinitely by means of an electric motor was devised. Today there are gyro-compasses of all sizes, used not only to steer ships but also to control the movement of aircraft and rockets.

The heavy gyroscope wheel spins on edge. There may be a U-tube filled with water fixed to one of the rings on which the spinning wheel is mounted so that when the curve of the earth would tend to cause the system to tip, the weight of the water running into the lower leg of the U-tube pushes against it and causes it to twist round instead. If the spinning wheel is lined up north and south when it is started, this arrangement will keep it always edgeways on north and south.

The gyro-compass which emerged from Jean Foucault's experiments is better than a magnetic compass because it is not distracted by the metal parts of ships (and, later, aeroplanes).

It did not take much further ingenuity to couple it up to a motor attached to the rudder and thus produce an automatic pilot. This can steer a ship or navigate a plane a great deal better than a human pilot can. For that matter, it can take a rocket up to the moon, round it and back again, because the gyroscope wheel when it spins – just like the Foucault pendulum when it swings – is operating according to the laws of mechanics in space. These hold true whether the operation is down on earth (in one's bathroom, for example) or up in space round the dark side of the moon.

Gyroscopic behaviour is shown by all spinning bodies. However gyroscopes and gyro-compasses behave more dramatically than ordinary tops because the spinning wheel, which is the basis of their operation, is heavy, spins quickly and because it is free to tip and turn. Although he did not know it at the time, Jean Foucault, back in 1852, was providing at least half the explanation of why a boomerang comes back. One of the principal delights of scientific knowledge is that one never knows what use it is going to be.

The proper way to throw a boomerang is end on; that is to say, in a vertical plane, with the curved or hooked shape pointing forwards. It must be thrown with an overarm action something like a serve at tennis and needs to be given as much spin as possible in the direction of flight. Although a boomerang is thrown straight up and down edgewise, its two sides are not the same: one side is flat, the other has a curved shape which is similar to the aerodynamic curve on the top side of an aeroplane's wing. In fact, this curvature on the boomerang serves the same purpose as the curvature on the wing of a plane.

This purpose is to provide lift. Because an aeroplane flies with its wings spread out horizontally and with their curved sides upwards, the direction of the lift as the plane moves forwards through the air is upwards. The faster it flies the more powerful is the lift. Similarly the slower it flies the less powerful the lift becomes until at last, if the pilot flies the plane too slowly, it will fall down. With a boomerang which is thrown edgewise, the lift is sideways. A right-handed man holds his boomerang with its curved edge towards him. Consequently he aims to the right of whatever it is he wants to hit, because as the boomerang flies through the air, the sideways lift of the curved surface pulls it to the left. But there is another effect involved which begins to explain why the boomerang comes back: because as it flies forwards it is also spinning end over end, the upper end travels faster through the air than the lower end. At the start of a good throw, a boomerang can be expected to travel at thirty metres per second or, to put this into more familiar English terms, 66 m.p.h. At the same time it is rotating end over end at about ten revolutions per second. Because of this, the tip that at any moment happens to be at the top is moving forward through the air at nearly fifty metres per second (that is 110 m.p.h.), while the tip at the bottom, which is in the process of coming backwards in the spin, is only advancing at ten metres per second, or 22 m.p.h. Since the amount of lift given by the curvature of the wing is proportional to the speed at which the wing goes through the air, it follows that the lift is greater for the top end of the wing than for the bottom. This causes the boomerang, which at the beginning of its flight is vertical, to tip over to the left. Because, as has already been pointed out, aerodynamic lift is exerted by the whole of its curved surface, the boomerang, while flying in the direction in which it is thrown, drifts sideways; but the top part drifts sideways faster than the bottom part, which is why it tends to tip.

But at this point a second principle comes into play, the principle of the gyroscope. When a man rides a bicycle and for one reason or another it tips to the left, if he wants to keep it upright, he soon learns (that is to say, his brain directs him) to steer to the left. The boomerang spinning through the air acts like a gyroscope – not being fixed by pivots or axles, it is free to obey the laws of dynamics. Just like a gyro-compass, therefore, when the aerodynamics of the curved side of its wing generate a force which causes it to tip, the gyroscope forces generated by its spinning oppose the tipping force by making it turn to the left. In fact, this gyroscopic effect operates like the brain of a man on a bicycle.

Although people in various parts of the world are reported to throw boomerangs, the original inhabitants of Australia are most famous for using them. Not all of their boomerangs are designed to come back. The first to be seen by travellers to New South Wales did so. The tribe employing it called it a 'bumarin'. But in Victoria, the native people called their boomerangs 'Marnwullun wunkun' when they were constructed so as to return but 'tootgundy wunken' when, after being thrown, they spun straight on through the air and did not turn back. The significance of the spinning throw combined with the special shape is that the aerodynamic forces generated keep the boomerang, whether of the returning or of the straight-on variety, up in the air and thus make it go farther. A good thrower can get a boomerang to travel about 200 yards through the air, either round a circular course which brings it back to him again (assuming he has missed the bird or kangaroo he was aiming at) or straight ahead.

As science is merely organized curiosity, it is not surprising that scientists do not change much from one century to the next. Jean Foucault back in 1850 had nothing to gain from watching the swing of a heavy ball slung from the ceiling forty feet up. Just the same, it was delightful to find out that such a pendulum could show, in its independent way, that it went on swinging regularly backwards and forwards while the ceiling, the room and the furniture, together with the scientist who was watching the whole business, were spinning round on a revolving earth. More than a hundred years later Professor Shapiro in Boston, Dr Binnie in Cambridge and all those Australians in Sydney fell under the same spell of that brand of educated curiosity we call science when they started to look more closely at the bath as the water ran away. So far, none of them has got anything out of it except the pure delight of knowing what happens. It is true that, just as Foucault's experiments with his spinning gyroscope eventually made possible gyro-compasses for ships and aircraft and guidance systems for rockets, so there is also a chance (as some people believe) that knowledge of the spin of water running down the plughole may lead to understanding, and perhaps control, of cyclones. These also spin but they spin upwards whereas bathwater spins downwards.

In the light of all this, no sensible person would grudge the time and money spent by Dr Peter Musgrove[1] and his students at the University of Reading in constructing a mechanical boomerang-throwing machine so that the precise way boomerangs fly can be studied scientifically without the inevitable uncertainties arising from the uses of human boomerang-throwers. Dr Musgrove suggests that the facts that this research turns up could perhaps make it possible to construct improved ventilating fans for central-heating plants or even some sort of specially efficient windmill to help solve the energy crisis.

As for myself, I don't care whether boomerang research leads to anything useful or not. But it's a delight to know that a boomerang in flight combines the mechanics of the gyroscope, the aerodynamics of a plane's wing and Newton's laws of motion, and that an Australian Aborigine, one of the oldest races on earth, intuitively mingled all these together to produce a weapon with which he could bring down a bird on the wing long before the University of Reading and its mechanical thrower were thought of. Yes, it is indeed delightful to know.

1 Musgrove, P., *New Scientist*, 24 Jan. 1974, 186.

8 Thixotropic, my dear Watson!

Generations of people have been puzzled, and many of the more tidy-minded of them worried as well, at why it is tomato ketchup does one of two things. Either it won't come out of the bottle at all or, on the other hand, it comes out in a large dollop. The more thoughtful members of this worried and puzzled congregation may also have been struck by the curious contradiction that the ketchup which a moment before stayed still, stiff as a jelly, unwilling to move or flow from the bottle, became entirely liquid and malleable once it was out of the bottle and flowing freely among the pieces of fish and chips on the supper-eater's plate. The answers to these questions are by no means trivial. They are intricate, useful and, at the same time, illustrate what a subtly organized, difficult, and delightful world it is we live in.

The key word is thixotropy or, to put it another way, tomato ketchup – like a whole lot of things (but unlike a whole lot of others) – is *thixotropic*. This means that there are times when it is difficult to know how to answer the question: is it a liquid or is it not? To start with, what do we mean by a liquid? The answer is that a liquid is any sort of stuff which, when you pour it, flows. It need not flow very fast, but it must flow if it expects to be called a liquid at all. Golden syrup, for example, *is* a liquid. On a cold day it will still flow out of its tin if the tin is tipped on its side. It will take some time to do it, but flow it will. On the other hand, take a bottle of ketchup from the shelf, unscrew the stopper, turn the bottle upside down and stand it on its head – and it does not flow. Must we take it from this that it is not a liquid?

Before answering this question, let the questioner take up the bottle, replace the stopper and vigorously shake the bottle – but *vigorously* and in the style of a barman shaking a cocktail. Then when the stopper is removed, it will be found that the ketchup can readily be poured out of the bottle. At this stage, therefore, we must undoubtedly accept that ketchup *is* a liquid. On the other hand, if, after pouring out a small amount, one replaces the stopper, puts the bottle back on to the kitchen shelf and waits to examine it the next day, it will be found to have reverted to its previous stiff, unpourable state. Must we therefore accept that tomato ketchup is neither one thing nor another, neither a liquid nor a solid, or is it both?

The explanation is an interesting one; it *is* a liquid but one fitted with a particular kind of built-in collapsible support. In brief, it is *thixotropic*.

To explain, it is probably simplest to start by talking about solids. Ice is undoubtedly solid as will be confirmed by anyone who has fallen down on it. Looked at carefully, ice can readily be seen to possess a crystal structure. When moisture in the clouds freezes solid to form snowflakes, it forms into enchanting crystal forms. When liquid water is frozen, it too forms crystals as it becomes solid. The size and shape of these crystals is affected by the speed with which they are caused to form. Nevertheless, a slab of ice always possesses the characteristic crystal structure which is a reflection of the molecular structure of the two hydrogen atoms and the one oxygen atom, the H_2O which makes water what it is. What happens when this lump of ice is warmed and gradually turns to water is that 'mortar' holding the molecules of H_2O together in the pattern of the crystal structure of solid ice is gradually weakened. The orderly 'brickwork' crumbles, as it were, and in due time the cohesion of the whole thing is lost. From having been an unpourable patterned brick structure, when it was solid ice, it becomes a pourable mass of rubble, so to say, when it is water. Just the same, however, the vestiges of the crystal pattern remain. That it, water, although it is a fluid, does retain a certain crystal structure.

Both water and paraffin are liquids, yet as is obvious to anyone who has poured them, their structure is different. As I have just described, the structure of water depends on the molecular structure of the H_2O of which it is made up. Paraffin is in chemical terms made up of long flexible molecular strands. In fact, these strands are composed of twenty or more carbon atoms linked together in long chains. This is why paraffin when it is poured behaves rather like a bowl of spaghetti poured out on to a plate: there is a tendency for the whole lot to come snaking out in one slurp. The molecular chains of paraffin are, of course, very much smaller than anything that can be seen with the naked eye. They exist as such chains nevertheless. Alcohol again is different in its underlying structure from both water and paraffin. All three, however, are liquid and all of them will pour if they get the opportunity. Paraffin will pour slowest, water less slowly and alcohol quickest. Yet all of them, being liquids, will pour regardless of their underlying molecular structure. Tomato ketchup, however, is different. To start with, it is a liquid with solids mixed in with it. But, most important of all, some of the solids, which are partly suspended and partly dissolved in the liquid, possess the ability of forming a subsidiary structure – a sort of temporary scaffolding – within the liquid. This is why such liquids – of which tomato ketchup is one – become solid, supported and made rigid on this auxiliary scaffolding. When the ketchup is vigorously shaken, the mechanical work causes the scaffolding to break down and the solidified liquid becomes fluid. When the liquid is allowed to stand, however, the scaffolding falls together again and the liquid once more becomes stiff and unpourable. This reversible behaviour whereby a stiff compound when work is done on it, for example, it is shaken or stirred, becomes liquid but, after the work stops, the scaffolding fits itself together again and the liquid resumes its stiffness, is called *thixotropy*.

All sorts of common things are thixotropic. Mayonnaise, for example, becomes quite stiff if it is allowed to stand. When it is vigorously stirred, however, it becomes liquid and pourable. It is a curious piece of modern history that, in the development of so-called 'convenience foods' that will sell, food manufacturers have actually been to the length of inventing a pourable salad cream. This has sold remarkably well. The odd thing is that nobody knows beforehand what degree of 'convenience' people will like and are prepared to pay for. It is not, after all, much trouble to cut oneself a slice of bread, yet sliced loaves have proved enormously popular and few large-scale bakeries who failed to provide them would stay in business. Similarly, it is not very arduous work to get a dollop of mayonnaise out of a bottle although, since it is thixotropic like tomato ketchup, it may sometimes be useful to give the bottle a sharp shake in order to get the contents out. The fact remains, however, that pourable salad cream – with the thixotropy taken out of it – has been a commercial success.

Golden syrup, though sticky, is an ordinary liquid and pours as such; heather honey is thixotropic. This is due to the fact that the bees not only bring back nectar to the hive but also mix with it a significant proportion of pollen protein. This mixed with the sugary part of the honey forms the basis of the supplementary scaffolding which produces the thixotropic effect. Take a jar of heather honey, unscrew the top and take it off and then turn the jar upside down and the honey will remain, stiff and immovable, behaving as a solid, not a liquid. If, however, after taking the lid off the jar, one stirs the honey well with a spoon and then inverts the jar, the honey will pour gently and deliberately out; it has behaved as a liquid should.

The internal structure of liquids and their behaviour when they are poured and pumped have been studied by scientists for 300 years or more. The Egyptians produced a clock, the clepsydra, which was operated by the steady drip of water running out of an orifice. It did not keep very good time, to be sure, but the fact that it was any good at all was a testimonial to the consistency of the underlying structure of water as a liquid. The Romans built great aqueducts to bring water from one side of the country to another, while their domestic plumbing could put some of our modern arrangements to shame. But if instead of handling water they had been piping crude oil about the place, they might not have been so successful. This is because crude oil is thixotropic and water is not. So long as the oil continues flowing through a pipeline and the pumps continue to pump, all is well. If, however, there is a hold-up, and the pumps stop and the oil gets the chances to lie still in the pipes, thixotropy comes into its own and the oil sets like tomato ketchup. This means that considerable work has to be done on it to get it going again.

The problem to be faced in pumping oil through pipelines is an example of the disadvantages which may arise from the existence of thixotropy. There are, however, a number of circumstances when it can be useful and ingenious people have been very successful in making use of it. One of the most readily available materials which exhibit thixotropy and which make it easy and convenient to study is a particular kind of mud. This is a mixture of water and bentonite clay (named after Fort Benton in Wyoming near which its sticky behaviour was first discovered). Mix 4 per cent of the dry clay with water and you get a gooey mess. Shake the bottle and it becomes a liquid; let it stand and it reverts to what it was before. On the other hand, if you make a stronger mixture containing 6–7 per cent of the clay in water and let it stand, the whole thing becomes a jelly. One can turn the bottle upside down and the jelly stays where it is. But if the bottle is vigorously shaken, the mixture turns into a liquid. What one has done is to break down the clay scaffolding. When one leaves the liquid to settle, however, the clay scaffolding falls into place again and, lo and behold, one finds that a jelly has re-formed.

Once upon a time, people used to clean their teeth with toothpowder. This was a nuisance to use and has completely gone out of fashion. Nowadays, everyone uses toothpaste. If one squashes out too much on the brush, some of it will flow off on to one's fingers or even drop into the washbasin. On the other

'This means that considerable work has to be done on it to get it going again.'

hand, if by chance one leaves the tube with the stopper off lying on its side on a shelf in the bathroom cupboard, none of it flows out, no matter how long it is left. This is the reason that the toothpaste-manufacturing people make it thixotropic by mixing a proportion of mud with it – to be exact, bentonite clay.

Let us – at least if we are ladies – take mascara. In order to put it smoothly on to one's eyelashes, mascara must, at the time, be liquid. On the other hand, once it is on, there are several very strong disadvantages to its remaining liquid. For example, one blink and the lovely lady would not be able to open her eyes again if the upper lashes stuck to the lower ones. And even if they did not stick, it would be highly inconvenient to have the sticky black liquid coming off all over one's own face – and worse still if it came off all over somebody else's! What the manufacturers do, therefore, is add an ingredient – whether it is bentonite clay or some other substance with thixotropic properties – so that while the mixture is in its little tube, it is in the form of a solid jelly. When the user wants to make use of it, she works it briskly with the brush provided. This breaks down the thixotropic structure so that the mascara becomes liquid. Further work is put into it during the course of the brushing by which it is transferred to the eyelashes. But then the liquid on the lashes is left to itself and consequently in a moment or two it resumes its solid jelly-like structure when it no longer has any tendency to run off or get transferred to other passing objects. At the same time the main bulk of the mascara in its tube, also being allowed to stay still, unruffled and unstirred, gives up its liquid state and – like toothpaste – becomes solid again.

Thixotropy is one of the qualities that makes facecream the attractive product it is. It is solid in the jar from which it does not spill. But when a couple of fingertips-full are rubbed into the skin, the work turns it into a liquid. Then, when it is nicely spread into all those bumps, blotches and wrinkles and left to itself, it turns back into a solid so that the user does not need to deal with the 'wet look' which she might not wish to have. And just as facecream primarily intended as a cosmetic to make ladies look more beautiful takes advantage of the principle of thixotropy so also do those useful jellies that engineers and garage mechanics use to clean the grease and grime off their hands. Such products are also made to be solid in their tins. But when they are scooped up and rubbed on to a man's hand, they become liquid. The liquidity, however, only lasts while the jelly is being worked. As soon as the work stops, the substance goes back into its jelly-like state. On this occasion the thixotropic ingredient is not clay but is wax or resin. But the principle is the same.

The best-known and perhaps the most useful example of thixotropy is its employment in non-drip paint. Those who have suffered the inconvenience and mess of painting a difficult corner situated above their head will recall the annoyance of paint running on to the handle of the brush. Even the most punctilious of painters, that is one who has covered the surrounding environment with protective sheets, will have been put to the troublesome bother of cleaning off drops of paint that have dripped while the work was being done. To the older generation, the phenomenon of being able to prise the lid off a tin of non-drip paint and then invert the container while the contents, stiff as a jelly, remain where they are, still retains an element of delicious danger particularly when the experiment is completed by replacing the lid, vigorously shaking the tin and seeing, when the lid is again removed and the vessel turned upside down, all the paint pour out.

In brief, the incorporation of a thixotropic ingredient in paint, whether it be clay or resin or any of a dozen synthetic compounds, has revolutionized painting and robbed it of half its difficulties – particularly when undertaken by amateurs.

Scientists are to be congratulated at having recognized that liquids possess a structure of their own, even though the kind of structure they possess in the moving, fluid conditions in which they exist is different from the more easily recognized structure of a solid. Not that the structure of some solids is not sometimes quite difficult to understand. For example, cotton, which is largely composed of cellulose, possesses a very definite structure which can be vividly shown up by taking X-ray photographs of it. Glass fibre, however, possesses something quite different even though one can make curtains out of it. Glass, indeed, is not (in scientific terms) a solid at all; it is a liquid. If one goes into a cathedral where there is medieval glass in some of the windows, it is often possible to see that over the course of centuries a piece of glass has actually 'flowed' as if it had been a sheet of dough which had been hung up by its top edge. A close look will show that the upper part of a particular pane from a window has become thinner from the glass gradually, as time has elapsed, having moved down to thicken the lower part.

Difficult though it may be to distinguish between what must be accepted as a stiff liquid and a soft liquid, it is even more difficult to work out and understand the behaviour of some of the liquids and solids as they exist in real life. Non-drip paint is an elegant mixture and a credit to the scientists who have put together what is often quite a complex collection of ingredients. The mathematics of thixotropy are in themselves formidably complicated. Yet how much more complicated are some of what we might take to be the ordinary, commonplace, simple liquids of daily life. Dough, for example, can be taken to be a thixotropic liquid. If one makes what a baker would call a 'slack' dough and leaves it on a mixing board it will start to flow, as a liquid should. Then, because it is thixotropic, it will stiffen up and become a solid which, if it is worked, will become liquid again. But at the same time that this is happening, the structure of the gluten fibres (gluten being a protein occurring in dough) will undergo an actual chemical change due to the process that engineers call 'cold drawing'. This is why, while dough must be worked enough if it is to be made into good bread, it must not be worked too much.

Blood, too, is a thixotropic liquid. But it is more complex in its behaviour than an 'ordinary' thixotropic liquid such as one of the muds of different thickness made up of bentonite clay and water. When one keeps blood protected from air, it can be made to behave as we should now expect it to. Like crude oil in a pipeline, it stays liquid as long as it is on the move. Luckily, except under drastic circumstances, it always keeps on the move. But if it slows up and stops, should there be an accident or injury, it tends to thicken. The ingredients which make it thixotropic are, however, almost commonplace. The machinery by which it clots to stop bleeding and produce a protective scab are quite separate. A whole chain of changes is involved of which, for example, the presence of calcium is one.

Liquids seem to be simple – but they are not as simple as they seem. Solids can be simple or complex and sometimes it is not easy to know which is liquid and which is solid. On top of the structural uncertainties there now comes the curious and unexpected state of thixotropy, used to such useful effect in toothpaste, non-drip paint and who knows what; a way to make neither one thing nor another. But even beyond all this are the *real* stuffs; the good cook with her 'feel' for her ingredients can cope with a problem of physics which a scientist cannot do properly *yet*. Tomato ketchup is rightly considered to be an admirable condiment and its physics, popularly enshrined in the deathless verse

You bang and bump and shake the bottle
First nothing'll come – then a lottle

does credit to its discoverer. Yet in life, many fluids are even more complex. Thixotropy is not the *last* word. It is a good word, however, which must always be remembered by scientists and non-scientists alike. There is more to be discovered and the discovery could still be either amateur or professional.

9 Smells like rain

It was someone in Australia who first thought of asking a scientist the question, 'What is the special smell that one smells after a long spell of dry hot weather when the first big drops of an oncoming shower start to patter down on to the dusty earth?' In other words, 'What is the smell of rain?' The scientists did not know; so they took the trouble to find out.

The first thing to be discovered was that the smell which we all at one time or another have smelt on a hot summer day is *not* the smell of dead insects or decaying vegetation. There is, indeed, a real smell of rain. And the next thing the scientists discovered, as they so often do when they start to study an unfamiliar subject, is that although ordinary people may not have given the smell a scientific name or worked out the chemistry of the stuff, its existence has been known about for a long time. In fact, there is a small but thriving perfume industry centred on a place called Kanauj about eighty miles from Lucknow in the Indian province of Uttar Pradesh which extracts, bottles and sells this particular smell under the name *Matti ha attar*. This name when loosely translated means 'earth scent'. It is odd that it has never caught on in Great Britain where the weather forms so persistent a topic of conversation. There is something rather charming – almost patriotic – in the thought of a girl putting a couple of drops of *Matti ha attar* behind her ears so that later on the young man making eager advances comes to a sudden stop, sniffs and at once remarks, 'Smells like rain.'

The way the scent is manufactured is curiously simple. In May and June they put out a number of clay discs where they are exposed to the hot sun. Well before the rains are expected, the men bring them in, break them up into quite small pieces, put them in a pot and steam them. The steam passes out of the pot through a tube leading into a jar of sandalwood oil in such a way that the steam bubbles into the oil carrying the scent from the clay with it. This peculiar scent attracted the attention of Australian scientists in the Division of Mineral Chemistry of the Commonwealth Scientific Institute of Research Organization, commonly known as the CSIRO. Just like everyone else they had in their time smelt the strange but pleasant smell one gets when it starts to rain after a dry spell – and it is not only people who find the smell exciting: horses and elephants as well become restless when they smell rain coming. And just like everyone else the Australian scientists had no idea what the smell was due to and what chemical compound it could be that the Indian scentmakers extracted from their clay bricks. Even so, the Australians gave it a name. They called it 'petrichor', which, when translated, means 'essence of stone'.

As a matter of fact the Australian researchers, Dr I. J. Bear and Dr G. R. Thomas, were by no means the first to have tried to find out what the smell of rain was due to. As far back as 1891 a very distinguished French professor called Barthelot came to the conclusion that it came from the remains of dead plants and he called it 'argillaceous odour'. As I have already said, this was incorrect. The bricks the Indian scentmakers use can be fired in a kiln which would destroy any decaying plant remains. Then in 1928 a Polish scientist got the idea that the smell was due to bacteria, whereas a later investigator writing in 1933 thought that the smell was due to fungus floating in from the atmosphere and landing on the earth.

111

It was the Australians, however, who showed that these early ideas were incorrect. They also did experiments which indicated how very strange the whole business is. For instance, whether one is in England, India, America or Australia, the smell of rain after long hot days is the same. It can, therefore, hardly come from plants because they are different in various parts of the world. Dr Bear and Dr Thomas also showed in their experiments that whether they spread out clinkers from the core of an extinct volcano, whether they used gravel, whether it was stone from a quarry or material from the spoil dump of a mine – all of these when roasted and then spread out in the hot Australian sunshine for anything between a month and a year, always yielded the selfsame smell of rain in the few drops of oily material that was finally extracted.

Even today scientists are not *quite* sure that they know what petrichor is: this strong romantic stuff possessing the smell of rain. Of course it has been analysed. The oily drops that are extracted from the clay bricks or the earth or clinker or whatever is used contain a mixture of substances. There is a lactone, an aldehyde, some nitro-phenols and several kinds of acid in it; one – called nonanoic acid – has nine carbon atoms. But in spite of all their efforts, the scientific investigators have not yet succeeded in determining *exactly* what petrichor is. But while all this was going on, another question had been asked – this time in America – which, though it sounded simple turned out to be remarkably difficult. The question was, 'What causes the blue haze that shimmers over the countryside on a hot day?' This does not sound as if it is a particularly difficult question to answer until one tries to do so. It then transpired that no one had any clear idea of what it was that produced the haze. Naturally there were scientists who then settled down to try to find out. In 1960 they published a learned report of their conclusions. Because this did not entirely answer the question, further studies were carried out. The results of these were reported in 1971.

In recent years we have heard a great deal about air pollution but even before this there were a number of regulations designed to keep the air we breathe pure. There has actually been a Clean Air Act, combined with restrictions on the use of anything other than smokeless fuel. It has been remarkably successful, so much so that 'pea soup' fogs have become a thing of the past. The notion is that not only is it possible to produce 'pure' air, absolutely and completely free from any 'pollution' whatever, but that air of this level of purity ought to be provided. The Americans studying summer haze now proceeded to demonstrate and publish their findings, this time in the *Proceedings of the U.S. National Academy of Science.* What they had found was that fresh air, collected on a summer evening when the haze was blue, was not just air but air with a whole lot of things mixed up in it.

Fresh air may be fresh but it is not just air. When bright sunlight shines into a darkened room when the blinds are down, the shafts of light streaming through the cracks can be seen to be dancing with motes of dust. Outside in the garden, as people who suffer from hay fever know all too well, floating among the motes of dust there are pollen grains. And in the air too are all sorts of other gases, besides the oxygen and nitrogen of which 'pure' air is composed. To start with, there is moisture or, to put this in scientific terms, water vapour. There is also carbon dioxide gas. In fact it is the carbon part of this carbon dioxide gas from which the skeleton of the trees of the forest and, indeed, all the plants that form our food, as well as those that do not, are made. Lie under a great beech tree and look up at the firm, smooth, grey trunk and remember that all the carbon in it was at one time floating about as gas in the atmosphere. It is odd to know that while carbon dioxide gas is a proper constituent of fresh air, one reason why the atmosphere of a stuffy crowded room is so stuffy is because of the carbon dioxide in it – this time there is too much for our comfort. So is carbon dioxide a *good thing* to have in one's air or pollution? It all depends on the amount.

Air naturally contains a small amount of what are called the rare 'noble' gases: argon, neon, xenon and krypton. Not much, indeed not enough to matter unless one is in the liquid air business in which case they are a nuisance. But in addition to water vapour, carbon dioxide, a little hydrogen and small traces of the noble gases – all of which were well known as being ordinary components of air – the research scientists investigating the blue haze of a warm summer evening were able to identify and measure substances which up till then had only been detected by smell. It should be said at this point that the sensation of the warm sun shining on the back of one's hand is produced by the electromagnetic energy of the sunshine. No material stuff has to strike the skin. Similarly with hearing, the vibration of the air causes the inner part of one's ear to vibrate and the vibrations are recorded by the brain. Again there is no actual stuff that one could identify and bottle as, say, the moo of a cow. Smell, however, is different: when a girl smells a rose, an actual chemical compound – a material called demascinone – gives off a vapour of itself and it is the molecules of this which, floating through the air, are actually sniffed into her nose. There they must settle on the sense organ where the shape and structure of the molecule causes the message that goes to the brain not only to indicate that there is a smell to be smelled but what kind of smell it is. The reason why we do not possess more information than we do about the nature of the compounds that cause different smells and the kind of chemical reaction that goes on in one's nose before the message is clearly sent to the brain (for example, 'this is peppermint' or even 'this is the vintage port of 1923') is that the amount of the smell-producing substance needed to produce the smell is exceedingly little. Today, however, there are exceedingly sensitive methods of analysis. And it was with the use of this sort of equipment that the people studying the blue haze detected a number of interesting substances.

There was a long list of substances which were identified as the volatile essences from leaves and grass. In particular, the material linaloon, which is known to be part of all sorts of plants but has been particularly considered as being a component of juniper berries was found. Since this is a characteristic component of the aroma of gin it would be more accurate when expressing enthusiasm about lovely weather, rather than saying that the air is like champagne, to describe it as being like gin.

The main point that has now been discovered, however, is that the summer air is full of these invisible substances that have come as vapour from the leaves and flowers of the forest.

'. . . the summer air is full of these invisible substances . . .'

The researchers who found this out went further: they calculated *how much* of all these compounds, each of which, remember, is only present in a very low percentage, was present (in tons) if the entire atmosphere of the earth were taken as a whole. The proportion of these volatile compounds in a cubic foot of air may be low but there are a great many cubic feet. In fact the estimate of the total weight of these invisible stuffs released from herbage into the worlds atmosphere in a single year was 438 million tons. This calculation was quite difficult to make because more of the volatile compounds float up into the air in summer, when the weather is warmer and when there are plenty of leaves, than in winter when it is colder and there may not be any leaves at all. Furthermore, in calculating for the world as a whole one must always remember that when it is spring or summer somewhere, it is, at the identical moment, autumn or winter somewhere else.

Let us keep all this in mind and before returning to the smell-of-rain question discuss another matter which also has a bearing on the subject. Kaolin, which is powdered earth originally derived from rocks rather like granite, is one of the principal components of toothpaste where its purpose is to absorb the general grime one gets on one's teeth. Kaolin is also used by people who have a queasy stomach; again, it acts by absorbing gas and perhaps the odd toxic component of spoiled food.

Since ancient times clays have been used in a similar way as absorbents – almost like mineral blotting paper. The Bible refers to one's sins becoming as white as wool (that is, after the wool has been treated with fuller's-earth, which has an ability, as it were, to blot up the greasy dirt on the raw wool). 'Claying' was a process used as long ago as the fourteenth century to make white sugar-loaves. What is now being discovered is that the clay of the earth, when it is baked dry by the sun, absorbs the pinene, myrcene and isoprene, which besides linaloon, have also been identified floating about in the atmosphere together with a whole lot of other compounds discovered by Dr Bear and Dr Thomas. All these are oils and essences derived from seed-pods, waxes of leaf surfaces, pine trees and many more. The conclusion reached by these Australian researchers was that these are the components that go to make up petrichor, the smell of rain falling on hot dry earth. All of them came originally from the summer atmosphere.

It is strange how wrong old ideas about fresh air can be. When I was a small boy, we were told that the special smell of fresh air at the seaside and the benefit it did to one's health were due to the ozone in it – without any strong scientific evidence, so far as I am aware, that there is any more ozone in sea-breezes than in other kinds of air or that it would do one any good if there were. As a matter of fact the particular smell that I was told as a child was ozone was, as now in my maturer years I am convinced, merely that of decaying seaweed. Next came the period, which is hardly yet over, that air to be really fresh, healthy and unpolluted, ought to have nothing whatever in it other than air. This of course poses a difficult problem, namely that of deciding what air is. Obviously it is more than oxygen and nitrogen, but how much moisture and carbon dioxide and argon, neon, krypton, xenon and half a dozen other gases commonly found in it it ought to contain is an increasingly difficult matter to decide. And what are we to do now that Dr Bear and Dr Thomas have found it to be full – or, at least, if not full, then peppered with – all sorts of things which, when they become stuck on clay and wetted off again possess this delightful smell of rain?

Scientists do their best to understand what is going on in nature but if science proves anything, it is that – in spite of electricity working the light, atomic bombs actually going off, and penicillin keeping people alive who without it would die – science can never prove that we know it all or, to use the language of lawyers, though what science throws up may be the truth, it can never be taken to be the whole truth. Obviously air pollution, to take it as an example, is a bad thing. But air pollution is not just a matter of there being chemical compounds floating about in the atmosphere. Even when there are 438 million tons' worth of them a year, they may not be pollution at all; on the contrary, they may be agreeable – or even useful.

The Australian scientists I have been talking about published a long and detailed paper in the scientific journal, *Geochimica et Cosmochimica Acta*, in which they not only put down their results but also did their best to speculate on what the results meant. To start with they pointed out that it was not merely useless curiosity that started them off in pursuit of knowledge of what the smell of rain was due to. Or if it was, the knowledge that had come to light could be a major step forward in our understanding of what was going on around us. It seems, for example, that a variety of organic compounds, some of them so-called hydrocarbons, derived from plants are constantly being distilled into the atmosphere. It is some of these that are being absorbed by the earth during hot dry periods and small amounts of oil can, in fact, be recovered from the upper layers of soil and rocks. Although Dr Bear and Dr Thomas were cautious in what they wrote, what their evidence implied was that these compounds, distilled off plants and leaves year after year into the atmosphere and subsequently absorbed on to dry earth and sand, might well become through the long progress of time the petroleum for which we all so eagerly thirst to keep our industrial life going. If this turns out to be true, it would very neatly explain the abundance of oil in the hot sandy places of the world.

We are all accustomed to the idea that the mighty rivers of the earth, the Ganges, the Euphrates, the Danube and the Nile – and, for that matter, the Mississippi, River Plate and the Congo – all in the first place rose up invisibly as water vapour from the seas and were carried equally invisibly in the impalpable air. We learn all this at school where the marvel and delight of knowing – about where rain comes from as well as how we become wealthy through the chemical industry – are taught to children long before they have to concern themselves with inflation or whether the pound is going up or down. Our children will no doubt cope in an equally matter-of-fact way, if the notions of Dr Bear and Dr Thomas turn out to be true, with the idea that millions of tons of what one day will re-appear as crude oil may also be ruffling our hair in the breeze, equally invisible as the substance that will one day become rain.

How sad it is that there is a feeling abroad today that science is bad, materialistic, the evil spirit hidden in the heart of our machines, the source of noise and pollution. Once our grandfathers believed that it would enable us with reverence to elucidate the beautiful truth and complexity of nature. And they were quite right. It was good luck that a research, started officially by the Australian Government with the imagination and faith sufficient to allow two experienced and highly qualified scientists to study the delightful (and potentially useless) subject of the smell of rain, came up with the finding that the same principle underlies the origin of the smell of rain and the origin of petroleum. But perhaps the most interesting spin-off of the whole study was the conclusion that pure fresh air may not be pure at all but full of odds and ends no one had imagined to exist until they were discovered.

Ask a question about the smell of rain and you get an answer about the chemical compounds that evaporate off the waxes and gums, the cuticle and the juices of plants and in the end get 'blotted' out of the air and absorbed on to the clay of the earth. But if this question leads to the things that come off plants and land up on earth, let us in the next chapter consider an even odder question which, in the answering, is concerned with traces of things which start out in earth of different sorts and end up in plants.

10 The crock of gold

Why, one might ask oneself, should three Soviet scientists spend considerable time in a wine-cellar analysing different bottles of wine? It seems a mysterious thing to do. Yet here is the report published in a very respectable, even if not very well-known, scientific journal.[1] The three investigators were G. I. Beridze, G. R. Macharashvili and L. M. Mosulishvili. The experiments were described in detail: samples taken from all the different bottles they tested were evaporated to dryness, the dregs remaining were roasted in a furnace, the ashes put into sealed quartz tubes and irradiated with neutrons from a nuclear reactor. The purpose of this remarkable exercise was – one would never guess – to determine the tiny amount of gold in each cup of wine.

The amount of gold in the various kinds of wine examined – so ran the report – was from 0.030 to 0.758 microgramme per litre. This is not very much, a microgramme being only one-

1 *Rodiokrimirya*, 11, 1969, 726.

millionth part of a gramme. For those who are interested to know, the amount of gold in red wine was found to be greater than that in white wine. But even taking into account the current fall in the value of money, 0.758 microgramme per litre of wine is hardly worth the trouble of recovering as a means of strengthening the exchange rate of the rouble. Nor are three serious senior Soviet scientists likely to entertain sentimental notions of copying Cleopatra's fancy for dissolving pearls into wine to improve the richness of its taste. Their purpose, while more rational and utilitarian, was nevertheless remarkable and, in its way, romantic.

Wines from different places – from Italy, France, Hungary or, for that matter, from California or New South Wales, or indeed from individual vineyards in these places – possess, as is well known, characteristic bouquets and flavours. These differences are due to differences in chemical composition which themselves are derived partly from the yeasts and micro-organisms – the bugs, so to say – native to different localities, and partly from the nature of the soil on which the grapes grew. This was the point laid hold of by the Soviet chemists. A large wholesale wine merchant has in his warehouse, in the form of bottles of wine, representatives of the grapes grown in vineyards whose precise geographical situation is well authenticated. These grapes are themselves a reflection of the soil from which they came and the soil itself is derived in large measure from the rocks near by. What the three investigators were doing, therefore, was an exercise in biogeochemical prospecting. Wine rich in gold is not worth discovering for its own sake but rather as an indication that the territory upon which the grapes were grown might be worth prospecting. With a country as big as the Soviet Union to cover, why wait for the accidental discovery of a Klondike or a California when the search can be directed more rationally from the comfort of a well-stocked wine-cellar?

The technique of geobotanical prospecting, as it can also be called, has been developed quite deeply during recent years, particularly by Soviet workers. And gold is not the only metal sought by this means nor are grapes the only plants used as indicators. Other botanical species have been found to be equally useful. Some of them speak to the investigator in an even clearer and subtler way. For example, *Astragalus*, which is made up of something like a thousand different species of the pea family, has usually been thought to be a rather useless and uninteresting collection of plants. Most of them are low herbs or shrubby bushes, although one of them, gum tragacanth, is used commercially in pharmacy and also in the manufacture of adhesives. To the geobotanical prospector, however, *Astragalus* has proved to be a 'universal indicator'. This is because most of the species in the family possess the ability to collect from the soil the element, selenium. In excess – and the excess may be very little indeed – it is highly toxic and can cause the death of cattle; in small amounts it is a necessary trace nutrient and is good for you. *Astragalus* picks selenium out of the soil with such thirsty efficiency that in parts of the Great Plains of America, in north-west Queensland in Australia and in certain localities in Canada and Ireland where soils are particularly seleniferous the *Astragalus* plants absorb so much selenium so quickly that they become poisonous to any livestock that eat them.

It is from this apparently undesirable characteristic that perceptive scientific workers have gained advantage. Botanists have gone out to Colorado to collect *Astragalus* and, by analysing their leaves, mapped out the seleniferous rocks below. Selenium is usually found associated with workable concentrations of uranium and this, of course, is what the modern generation of utilitarian scientists is really after. Their predecessors who collected knowledge for its own sake did so because of their delight in discovering what the earth was made of. But there is satisfaction too in being able to use such knowledge to get the fuel to warm the houses and run the factories by which the conveniences of civilized life in the twentieth century are made possible.

Astragalus is not the only universal indicator. A species of viola, *Viola calaminaria*, is useful as an indicator of zinc. *Vesicaria alpina*, which possesses the ability to collect copper, has been used by Norwegian prospectors in their search for the metal. *Silene cobalticola* becomes rich in cobalt when it grows

'Selenium is usually found associated with workable concentrations of uranium . . .'

where cobalt is plentiful, and this ability has been used to search for cobalt in Katanga.

The plants I have so far mentioned pick up particular elements – copper, zinc, selenium or cobalt – where they are plentiful but flourish just as well where they are not. Other plants are used by miners searching for minerals and indicate their presence in a different way. First, there are the so-called 'local' plants. A sensitive prospector searching for nickel in Italy – or in Georgia, for that matter – would always, if he understood his job and knew his flowers, pay special attention to those places where he noticed the pretty blossom of *Alyssum bertholonii*. This plant is better adapted to territory where nickel ores occur and there it grows most plentifully. Again, in China they dig for copper where a particular kind of escholtzia blooms. Montana silver-miners know that where *Eriogonum ovalifolium* is to be found, there silver is likely to be found as well. And in Czechoslovakia it is known by those who understand where to look that *Equisetum arvense* favours soil on the hills where the rocks bear gold.

From all this it can be seen that there are two ways to use plants when searching for metal from the rocks below. One way is to analyse the leaves or fruit or even wine made from the fruit and dig below those plants whose tissues contain most of whatever one is looking for. Another approach is to search about until one sees a clump of 'local' plants, the very presence of which is an indication that one metal or another is likely to be there. But there is a third way in which a plant can be useful to a miner: there are some plants that indicate the likely presence of rich ore by *not* growing near by or, if they do grow, coming up stunted or deformed. For example, the kind of anemone called *Pulsatilla patens* is common over wide areas of the countryside in the Soviet Union. When, however, a quick-witted Russian prospector comes to a place where none of these anemones grow or, if they do, where their flowers have lost their petals and their leaves are small, he stops and looks for nickel. The reason is that *Pulsatilla patens* is highly sensitive to nickel.

While it may be useful to Soviet mining engineers to know where to look for gold for money and for nickel to plate the handlebars of bicycles, the behaviour of the tell-tale plants themselves is interesting for itself alone and could indeed lead to a number of other useful possibilities. The 'indicator plants', for example, which are able to absorb so much of the particular metals they indicate without at the same time poisoning themselves raise the question as to how they do it without coming to harm. It is all very well for the prospectors to move on from one place to the next, picking the flowers and analysing them so as to be able to edge towards those growing right on top of a mineral deposit where the amount of nickel in the leaves and flowers may be very great indeed. But how, you may ask, do such plants survive? For example, when *Alyssum bertholonii* has got a really rich nickel deposit to indicate, it may absorb so much of the metal into itself that, when it is burnt up in a furnace, its ashes contain 10 per cent of pure nickel.

Under ordinary circumstances a concentration as great as this in a living plant would be highly toxic and the plant would die. Why doesn't it? The answer seems to be that these 'indicator plants' have, by the long process of evolution, developed the ability of taking the metal they do absorb out of circulation. They have inside them a complicated, big, organic molecule which combines with the metal and thus holds it still and prevents it from circulating round the living cells and doing them harm. This process has been called *biochemical segregation*. But not only do such plants possess a mechanism by which metals they suck up from the soil do them no harm, they may even come to thrive under circumstances that would be lethal to ordinary plants. There is at least one report in the scientific literature of plants which flourished all round a disused lead-mine where the soil contained 1 per cent of pure lead but grew badly when they were transplanted into what would normally be accepted as a good, fertile, lead-free soil.

The ability to link up in a complex chemical molecule what would otherwise be a toxic material is not unique to plants; animals can sometimes do it as well. Plaice, for example, possess a remarkable facility to pick up arsenic out of the sea. But it does not poison them because their cells have the capacity to bind the arsenic into a particular chemical configuration which is inert. By this means the arsenic is so effectively inactivated that not only does it do the fish no harm but neither does it harm you and me when we eat the fish. The only people who could be said to be inconvenienced by the arsenic in plaice are the officials whose business it is to draw up regulations about food safety. If they set the permitted maximum considered to be a safe concentration too low, perfectly wholesome foodstuffs such as plaice will be classified as unfit for human consumption. This would be quite wrong because, as I have said, although there is some arsenic in plaice it is tied up in a harmless form. On the other hand, if the standards are set too high, here and there someone will be poisoned. For example, oysters possess of facility for picking up the metal, zinc. In fact, when a number of them were submitted to analysis, it almost seemed as if, should the dry substance of the oysters be hit with a hammer, it would ring like a bell. Yet though the superabundance of metal seemed to do the oysters no harm, if the oysters were eaten by people, an instant reaction supervened and those who had partaken found themselves suffering from food poisoning.

People often ask a scientist what he is doing. Then, when he tells them – for example, that he is studying why it is plaice don't die of arsenic poisoning, or how it is rabbits can go on eating the flowering plant, Deadly Nightshade, to their hearts' content without suffering harm – they respond by saying, 'What good will it do when you find out?' The short answer to this question is that nobody can tell; the long answer is that the question itself is unimportant. Real scientists do research because they want to know, not because they want to get rich, or even to be benefactors of mankind. But what is particularly interesting is the fact that just as what was once thought to be useless knowledge can often later on – sometimes centuries later – turn out to be useful, so too may useful knowledge turn out to be interesting in an unexpected way. The three Soviet scientists may have been doing no more than their duty when they measured the amount of gold in one bottle after another of all the different Soviet wines they could lay their hands on. Yet out of this straightforward, even tedious work there came (quite apart from information about the best place to dig a gold-mine) new ideas about the way plants (and perhaps men as well) could protect themselves against poisons by 'complexing' the toxic agent inside them. That is to say, by causing it to combine with an organic substance in such a way as to form a compound which has the effect of taking the toxic agent out of circulation.

Shakespeare suggested that we might find 'tongues in trees, books in the running brooks, sermons in stones, and good in everything'. True though this may be in its poetical sense, scientists may conclude that in their work they are more likely to discover 'sermons in books, and stones in the running brooks'. Be that as it may, the discovery of gold in wine and the ability of *Agrostis tenuis* to resist high levels of the metal, lead, may preach a very direct sermon to people. The ability to 'wrap up' a poisonous substance and take it out of circulation may decide whether the creature that possesses it lives or dies.

Most of the protein one eats is used as fuel to keep one's body going. But whenever protein is so used, the nitrogen in it, which has no energy value, has to be excreted out of the body. This nitrogen ends up in the form of ammonia, strong-smelling and highly toxic. How then are we to eat protein without poisoning ourselves with the ammonia we produce? The answer is that we have over the ages acquired the knack of turning ammonia in our bodies into a more complex stuff, urea. This is a good deal less poisonous than ammonia. Too much of it is harmful but normal healthy people get rid of it in their urine. Unfortunately patients suffering from kidney failure can, however, die of poisoning from the build-up of urea in their blood. At this point I should like to refer to the question which intelligent people may be asking themselves about birds. As everyone knows, birds live a significant portion of their lives in a closed eggshell. How then, you may ask yourself, do they get rid of the urea which must be accumulating for the entire three weeks (assuming the birds we are talking about are chickens) during which the creature spends all its time closed in? No animal that excretes the breakdown products of the protein it consumes as urea could go three weeks without passing water and thus ridding itself of the urea which would otherwise poison it. The answer is that birds have evolved a chemical trick by which they convert the ammonia left over from the protein, not into water-soluble urea, which would face them with the problem of getting rid of the water and its urea, but into uric acid. Uric acid is quite insoluble and the developing chick can get rid of it as crystals which can stay tidily in the eggshell not doing anybody any harm until the chick is hatched and can get out.

This is one example in animals of taking a poisonous substance out of circulation, which, as we have seen, can be done with such diversity by plants. If we could find out more about how plants – and animals as well – do this, the discovery might be very useful in medicine.

11 The tip of the iceberg

Why do people talk about the tip of the iceberg when they mean a little bit of something showing, of which there is a great deal more concealed under the surface? Or – to put the question rather differently – why do icebergs, which are often as big as a fair-sized island or a complete city block, float so low in the water? Or – to ask the basic question underlying the whole business – since icebergs are, after all, only made of frozen water, why don't they merely take their place half-way down or, for that matter, on the bottom instead of on the top of the sea? All these are good questions and, like so many of those asked by non-scientists to scientists (or even by children, who often ask the best and most direct questions of all), they are not altogether easy to answer.

Perhaps before starting to answer it is worth pointing out that this world would be a very different place if ice did *not* float. Ponds, as every schoolboy knows, are full of wild life; and the oceans too, quite apart from the rich diversity of fish that live in them, are the home of a variety of plants and animals. If, when the temperature dropped, a pond started freezing from the bottom, many of the species of creatures that have their home there and have evolved a means of carrying on in a manageable environment, even during the hardest of winters, would be immobilized or, at best, gradually driven to the surface. Conditions for the marine animals would be even more destructive. As it is, though the Arctic night may shut in the seas and their inhabitants under a thick layer of ice many feet deep, underneath the living world goes on and, at the bottom, the temperature of the water never falls below 4°C (39°F).

There are two reasons why icebergs float at all, leaving aside for the moment why they float so low in the water. The first reason is that a lump of water, of whatever size, becomes, when it is cooled down, a lump of ice a size bigger. There are a lot of complicated scientific reasons why this is so. These are, indeed, so subtle and involved that we can feel some sympathy with the great nineteenth-century scientist, Count Rumford, who, believing that all other liquids on earth became *denser* (that is to say *heavier* per unit lump) when frozen than they were when liquid, thought that the reason why water alone became less dense (in fact lighter) when it froze was due to the miraculous interference of Providence. However, he was wrong – at least so far as water being the *only* liquid that swells when it freezes; there are, as it happens, one or two others, although not very many. But the best explanation that can be given of why water increases in size when it is frozen is that the way the three atoms comprising water (the two hydrogen atoms and the one of oxygen that make up H_2O) stick out and arrange themselves in relation to those of the other surrounding water molecules gives the solid ice an open crystal structure. There are in fact several varieties of this structure but all of them produce ice that is bulkier than the water it comes from. It is actually possible to make a heavy ice that sinks instead of floating but this can only be done in a laboratory by subjecting ordinary ice to enormous hydrostatic pressure.

The straight fact of the matter is that when the water freezes the H_2O molecules milling about in liquid water fit themselves together to form ice with a crystalline framework in which the molecules occupy a little more than 8 per cent more space than the water did. This increase in volume is enough to burst an unprotected pipe which happens to be full of water on a night cold enough to freeze it. And if we go back to the iceberg, it is easy to see that because ice is lighter, volume for volume, than the water around it, an iceberg floats. But because icebergs are only 8–9 per cent lighter – very much less than ships, which contain a large volume of air – they float low in the water.

There is another reason why icebergs float instead of sinking and it is this: sea water is heavier, volume for volume, than fresh water, because the water in the sea is salty. It follows, therefore, that, whether or not it was turned into ice, a volume of fresh water, provided one could prevent its mixing, would float on the top of a mass of salt water such as is found in the sea. One could, in fact, do an experiment to show this by filling a plastic bag full of fresh water from the tap and placing it in the sea. The bag of water will float. When the sea freezes and the top of it turns to ice (some of which in due time may break away and drift off as an iceberg), the ice that forms is fresh water; the salt is left behind in the water underneath.

Salt sea water is curious stuff. For the most part it is a 'slippery' pile of tiny molecules of H_2O moving about like a crowd going to a football match. But while most of the crowd milling about are in threes, as it were, with O (the oxygen) holding on to two Hs (hydrogen) to make up H_2O, some of them are in pairs, OH, while some of the hydrogens are on their own. In sea water, mixing with this crowd are a few strangers. These are molecules of the different kinds of salt which combine to make sea water salty. When the sea begins to freeze, the Hs and the OHs gradually stop moving about and get together, like the supporters of the home team at a football match. The Os and the Hs then start to attach themselves in a systematic order to each other and, as I described above, form up in the strong but open structure that takes up more room than before. This operation squeezes out the 'strangers' – the supporters of the visiting team, so to say – that is, the molecules of the various substances making up the sea salt.

*

There are two reasons why practical people, and scientists as well, have in recent times been paying a good deal of attention to the tip of the iceberg and the much greater mass of ice that lies underneath it. Long before 15 April 1912, when the *Titanic*, the largest liner then afloat, on her maiden voyage across the Atlantic went down with the loss of 1,513 souls out of 2,224 on board, navigators had been alive to the danger of icebergs to ships. Nowadays, when ships are equipped with radar and in addition can be warned of the danger of obstruction by these great floating mountains, they can take avoiding action. There is, however, one kind of maritime structure which cannot alter course when icebergs bear down towards it: the off-shore oil-rigs, each one bigger and more expensive than the Eiffel Tower. Because of this, scientists have set themselves to study seriously the various factors that make icebergs sail about the seas the way they do.

It costs about £15,000 a day to drill for oil in the North Sea, quite apart from the cost of the great steel tower – the rig – from which the drilling is done. If an iceberg suddenly appears on the horizon heading straight for the drilling platform, the oilmen are in trouble. And part of their trouble when they first began their North Sea operations was that they had no real idea how icebergs move. The first thing that came to light was that there are two main types of iceberg. Those that come from around the North Pole are what one might describe as orthodox icebergs. They are jagged and mountainous. In spite of the fact that the largest part of them is under water, some of these northern-type icebergs tower up from the water and present a frightening, if beautiful, sight as they approach. Sometimes as they move across the water they may suddenly tip or roll over altogether. Two forces make them move: ocean currents and wind. But they do not necessarily move straight downwind. As often as not, their craggy superstructure, derived from the un-even northern glaciers from which they have broken away, act like sails. With the wind in a certain direction they tack and veer along an unpredictable zigzag course.

The first notion of how to prevent these great things crashing into the oil-drilling platforms and destroying them was to blow them up. Experience soon showed that this was not a practical proposition. Some of the icebergs were miles long and consti-tuted veritable floating islands of ice. The destruction of these monster structures having been found to be impracticable, an alternative proposition was tested – on the face of it, one would have imagined, unlikely to succeed but nevertheless worthy of examination. This was the notion that the drifting icebergs might perhaps be towed out of the way. Cargo ships, notably oil-tankers, have already been constructed to an enormous size and steered around the world without undue difficulty. Surely, therefore, it might be possible to move icebergs in the compara-tively open waters of the northern seas? And, in fact, it can be done. The power developed by big ocean-going tugs is now very great; the development of light, yet strong, cable made of nylon and the like makes it possible to exert the necessary force to get the ice masses in motion. Finally, while it may not be possible to destroy the larger icebergs entirely by explosives, it has been found feasible to modify their shapes by splitting off parts here and there; explosive charges are drilled in at appropriate points and set off in a controlled manner. In the same way that quarry-men bring down parts of a cliff to allow themselves to get at the mineral products they wish to handle, tugmen can modify the outline of their icebergs and make them more manageable.

Having succeeded in large measure in developing a means by which icebergs could be steered about the sea, the people con-cerned in such operations began to become more ambitious. If it is possible, they thought, to direct icebergs out of the way and send them off into the south Atlantic to melt, might it not be possible similarly to pull other icebergs about the world and set them down where they could be useful? In their attempts to do this they turned their attention to the second kind of ice-bergs. These are icebergs of the 'table' or 'platform' type which are formed on the shores of the Antarctic continent.

Why should anyone want to tow an iceberg, you may ask, and what good can it be to the people who tow it when they reach their destination? The answer is that fresh water is a valuable commodity, that people like to use more of it as their standards of living rise and, perhaps most important of all, one of the main factors limiting the production of food, particularly in hot and arid regions of the world, is a shortage of water for irrigation. All sorts of means have been tried to increase supplies of fresh water: deep wells have been dug, aqueducts have been built from distant times past right up to the present day, and schemes have been put into operation for desalination of sea water – that is, distillation processes to separate fresh water from the salt water of the sea. Since the great ice mountains which are icebergs are already made of fresh water, might it be possible, the question was asked, to direct them where fresh water is wanted and where today people are paying substantial sums of money to get it by one or other of the various methods I have just described?

It is interesting to note that this idea is not new. As long ago as 1853 a ship which was chartered to bring ice to San Francisco all the way from a particular lake in Alaska, being unable to obtain a satisfactory supply, loaded up with glacier ice which, had the shippers left it alone, would have broken off into the sea and sailed away as an iceberg. And if this is not exactly what we have been talking about, it can be noted that a little later, between 1890 and 1900, groups of men succeeded in fitting small flat icebergs from the Antarctic with sails and actually sailing them to Valparaiso and even to Calloa in Peru, a distance of more than 2,400 miles. Other seamen towed icebergs by ship the same journey. The modern story, using the super-tugs of today is, however, more remarkable still.

After surveying Antarctica to locate a reliable source of a regular supply of large-scale icebergs, the latter-day seekers after water have found two particularly attractive areas. These are the Amory Ice Shelf on the east coast of the Antarctic continent and the Ross Ice Shelf on the south coast. Of all the icebergs floating off from these two zones, most rose 120 feet up above the waterline and were about 500 yards long; many rose 250 feet from the water and were more than *two miles* long. The biggest rose 350 feet high – this takes no account of the amount of ice *under* the water – and were four miles long. When proper studies had been made and such characteristics as the *drag* factor had been worked out, it was calculated that large modern tugs could tow these monsters economically at an average speed of one knot. Of course, the power required to tow an iceberg is greater at the beginning of the journey when there is more ice to pull. But it is not merely a matter of the tonnage of ice; the shape of the iceberg has a bearing on the matter. As it happens, the amount of melting is greatest at the front part so that as the iceberg is hauled along, it gradually acquires a more streamlined contour.

Considering the distance that has to be covered before one can sell an iceberg for the fresh water it contains, it is remarkable how much of it survives as ice to the end of the journey. Obviously, the bigger the iceberg the more of it will be there at the journey's end. Careful studies have been made for two journeys. One is from the Amory Ice Shelf in Antarctica to Western Australia, a journey of 4,700 miles, and the other from the Ross Sea to the Atacama Desert in South America which is 5,000 miles. As long as the iceberg is bigger than, say, one mile across, about 70 per cent of it will still be in the form of ice when it gets to the South American desert, while 75 per cent or more will reach Australia.

The objection might be raised that although it is all very well to tow an iceberg from the frozen wastes of the South Pole to the arid regions of Western Australia or South America to serve as a supply of fresh water, how does one pipe it into the water supply when one arrives? The best solution so far put forward has been that of an American called Isaacs: that when an iceberg arrives off the coast and has been satisfactorily moored, it should then be surrounded with a floating fence-like baffle, arranged so as to extend some distance under the water. When this is done, the fresh water which trickles down the sides of the iceberg as it melts does not mix with the heavier salt water of the sea but forms a separate layer on the top, which can be pumped ashore.

The actual amount of fresh water obtainable from an iceberg supply is considerably more than what is contained in the ice that arrives, and there is a good scientific reason for it. The explanation is simply that the ice is cold. In the hot and arid coastlands where water is most urgently needed it often happens that at night and in the early morning dense fogs roll in from the sea. This is particularly true along the west coast of South America where the cool Humboldt Current runs northwards in the Pacific. As long as an iceberg is moored off shore, these fogs condense on its cold surface and the water thus formed trickles down to mingle with that of the slowly melting ice. This supplementary supply of fresh water may be quite considerable. The actual amount will depend on how quickly the iceberg melts and is pumped away. The slower this happens the larger will be the bonus of water condensed on the spot.

If the ice-water is to be used for irrigating farmland rather than for drinking, there is another way in which its volume can be increased. The water supplied by an iceberg is extremely pure, far purer, for example, than 'hard' London water. As a general rule there will be less than ten parts of impurity in a million parts of water. But water for watering crops need not be as pure as this. It is good enough to sprinkle on the land even if there are 400 parts of impurity (say, salt) in each million parts of water. Consequently, for irrigation it is perfectly legitimate to add a certain amount of sea water to the iceberg water. And if the crop to be grown is cotton, the proportion of salt can be as high as 1,500 parts per million. This allows one to mix in as much as $4\frac{1}{2}$ per cent of sea water. It is interesting to note that the water in the lower reaches of the Colorado River naturally contains about 900 parts of impurity per million and yet can be used to irrigate most of the crops grown by the farmers along its banks.

The natural cooling effect which a captive iceberg towed to the coast of a hot dry land exerts on the atmosphere off shore, by which extra water is condensed out, will obviously be exerting an influence on the climate. In other words, it must make the on-shore breezes cooler. At the same time it will to some considerable degree lower the temperature of the sea. This local effect may be good or it may be bad. For example, when a power station, whether it is run on coal or on nuclear energy, is sited near the coast, it has a tendency to warm the water. This warming effect is sometimes called 'heat pollution', although holiday-makers bathing off the coasts of Britain would be inclined to describe anything which made the water warmer by some kindlier term. But where warming of the water might indeed damage the environment for the plants or the fish naturally found there, it could be advantageous to ensure that a

'. . . it has a tendency to warm the water.'

power station on shore is built opposite an iceberg mooring jetty off shore.

Although an iceberg towed away from its natural course to serve as a supply of fresh water in America or Australia would have some effect on the sea temperature and the climate at the position to which it was brought, there would be no effect on the world's climate as a whole. Already, without any intervention from scientists, water engineers and mammoth tugs, icebergs starting from near the Poles drift long distances from north to south into the warmer waters of the oceans. Each iceberg spreads around it in the atmosphere above a cloud of cooler air and in the sea water below a belt of cool, fresh water. If towing should be adopted as a regular operation, it would merely involve the same amounts of this cool air and fresh water being shifted to a different location.

Not long ago (in 1972, to be precise) quite careful calculations were made to see whether icebergs really did constitute a reasonably cheap source of fresh water. At that time a super-tug, approximately equivalent in power to a large ice-breaker, cost $60 million. Assuming the tug would last twenty-five years, this sum is equivalent to $2.4 million a year. To this must be added the wages of a sixty-man crew of about $2.1 million a year and maintenance, repairs and interest charges coming to, say, another $4 million. On top of this come the costs of fuel for the slow pull from the Antarctic to where the ice is to be used and the quicker, easier journey back again. Depending on the size of iceberg towed, these calculations showed that by the time the ice had reached its destination, it could be expected to have cost something not very different from the cost of water produced in some of the large desalination plants that are already operating. So it is feasible that arid, desert stretches of the earth's surface could be irrigated to grow food and become green and pleasant lands where people would like to live – all because water when it freezes increases in bulk (unlike most other liquids), and because when water is salty, the ice that forms from it is not.

We talk about the 'tip of the iceberg' when we refer to the comparatively small amount of ice we see when ice is floating low in the water. We also mean that a small thing may be an indication of a greater complexity, if we had the wit to think and understand. Thinking is a curious activity and may lead to strange conclusions. My cousin thought about ice and arrived at a conclusion which, strange though it was, was sufficiently plausible to attract the serious attention of the Supreme Commander of Allied Forces in South-East Asia in 1942.

Geoffrey Pyke, at that time serving as an ideas man in a special unit which formed part of the staff of Lord Louis Mountbatten, was struck with two ideas about ice, neither of which we have so far discussed. The first was that ice is very hard: anyone who has ever gone skating and has fallen down knows this, it may be said. Yes, indeed, but while everyone knows that ice *is* hard, few people have bothered to measure how hard it is. Its hardness, as my cousin Geoffrey Pyke discovered when he had obtained the necessary information about it, is very similar to that of concrete. Consequently, if one constructed a structure of ice and reinforcing steel rods, as used in the building industry, one would obtain a substance with the strength and consistency of reinforced concrete. But reinforced concrete that would float.

Geoffrey Pyke's second idea was more remarkable in its way than the first. If one makes a slushy mixture of water and wood pulp, using a recipe with anything from 4 to 14 per cent of wood pulp to water, and freezes it, one obtains a substance that is not only from two to twelve times as strong as concrete but one which is quite difficult to melt. The idea was to use this material – which was given the name pycrete – to build giant aircraft carriers. These were to be 2,000 feet long and with hulls thirty feet thick. Though these dimensions might have seemed large for ships in 1942, they were small for icebergs. And the proposed pycrete vessels were to have been proper ships, yet in war they would have been just as invulnerable as icebergs are. Not only was it calculated that a torpedo striking one broadside on would only make a crater in its side three feet deep and twenty feet across, but even if it did, cold air from a refrigerating plant carried in the ship would be circulated through cardboard tubes fitted veinlike into the pycrete hull and, by adding the appropriate proportions of wood pulp and refreezing the required

139

amount of sea water, the crater could quickly be repaired.

In Geoffrey Pyke's memorandum to Lord Mountbatten the idea for constructing a pycrete ship was worked out in detail. The splendid thirty-feet-thick deck would make an admirable landing-place for aircraft. Down below, in the hangar where the aircraft were to be parked, a false door was to be installed so that the aeroplane need not stand about on the ice. And the same arrangements were planned for the crew's quarters: far from being cold, the living accommodation would have been particularly snug. Ice is an excellent insulator: anyone who has ever visited a colony of Eskimos is struck with the warmth of an igloo, where the men sit about in their undervests. It would have been even snugger in an aircraft-carrier made of ice because it was planned to line the accommodation with panelling, ceilings and floors. The whole idea was quite thoroughly researched. For example, a lot of work was done to assess the engineering characteristics of pycrete maintained at a number of different temperatures and such mechanical characteristics as 'creep' and 'fatigue' were studied. Indeed, had not the atomic bomb been dropped on Japan and the war come to an end, ice-ships – not crude icebergs but properly engineered and metal-clad vessels made enormously strong and especially buoyant – would almost certainly have appeared on the oceans of the world.

Geoffrey Pyke thought up a number of cunning ideas specially suited to warfare. For example, pycrete landing-craft would be excellently well fitted for amphibious operations. Approaching a hostile coast with the ramp – thirty feet thick, be it remembered – would provide admirable security for the attacking troops. On reaching their target area, the commando forces, with a few swift strokes of a specially designed blow-torch applied along predetermined fracture lines, could cut out part of the side of the craft, like a drawbridge dropping in an ancient fortress, and out the armoured vehicles would roll. Similar advantages could be conceived in the design of pycrete oil-rig supply ships and cross-channel ferries.

Now that times are hard, we may perhaps see these ice-ships whether or not a war situation arises. Steel needs to be made and supplies of iron ore must one day become more difficult to obtain. There will always be plenty of sea water. It takes a great deal of energy to smelt iron ore and turn it into iron and then convert the iron into steel. The amount of energy to produce equivalent amounts of pycrete is very much less and would be even smaller if shipbuilding were established in north Greenland or even on the Antarctic continent. The possibility, curious though it may seem, has never been rigorously denied, and with this material of construction we must all agree that the possibility of scarcity does not exist.

12 The moon and Donald Duck

Anyone who saw the first astronauts wallowing about on the moon is unlikely to forget the strange feeling of eeriness at the sight. We are used to space travel now and hardly give a thought to how marvellous an achievement it is for human beings to have been able to get themselves up there, walk about in those great clumsy diving suits it is necessary to wear and – perhaps most remarkable of all – get back down again. A lot of the work and engineering and rocketry can hardly be described as science. It was technology of the most advanced kind – to store all the oxygen for breathing; to make the space-suits so that one could walk about in them; to design the big three-stage motors to get the rocket off the ground; to make all the little rockets to enable the pilots to steer the space modules; to design the radio telephones and television sets to give the fliers two-way communication with the ground station – all these were remarkable achievements even if they were not exactly new discoveries in science.

Of all the eerie qualities of the moon as a place to walk about on and explore, one of the strangest things, it always seems to me, is that it is absolutely silent. If the radio sets in their helmets went phut and if it was possible for the men to put their heads out to listen, they would not hear a sound even if one of the rocks fell off a moon mountain close by. The reason for this is that there is no atmosphere on the moon. On earth, you can hear a certain amount under water and if you put your ear to the ground you might be able to hear the thud of distant footsteps, but for the most part everything we hear is carried to our ears by the air. One of the simplest and most elegant experiments that teachers use to demonstrate the truth of this is to set an electric bell ringing. The sound is loud. The bell is then covered with a glass jar, something like an elaborate dish cover. The sound is less loud but it can easily be heard. But if next a pipe is attached to a nozzle on the jar and the other end connected to a vacuum pump, the sound gradually becomes fainter and fainter until, when a good vacuum has been established under the jar and almost all the air has been pumped out, not a sound can be heard even though one can see that the bell is still ringing.

Sound is transmitted by the trembling of the body of air that lies between whatever it is that is making the noise and one's ear. It is rather as if one had one's head in a swimming-poolful of jelly. The wobble imparted to one side of the jelly transmits the same motion to the other side. The note one hears depends on the frequency of the vibrations of the air striking one's ear. The quicker the beats come, that is to say, the more of them there are per second, the higher is the note one's ear transmits to one's brain. People's ears are most sensitive to sounds varying between low notes that beat the air at about 1,000 times a second and high notes that cause it to vibrate 4,000 times a second.

All sounds, no matter whether they are high notes or low ones, travel through the air at the same speed even though there is a quicker rate of vibrations from the high notes than from the low ones. The speed of all these different sounds has to be the same because if some notes travelled faster than others it would be impossible to listen to an orchestra or a brass band – or even a Welsh rugby crowd – from a distance without the music becoming all mixed up. On an ordinary day the sound of a band or, for that matter, of a man hitting a cricket ball, comes towards one at a speed of 767 m.p.h. This may seem to be fast but it is not in fact as fast as all that, as people who travel in ordinary aeroplanes, quite apart from those who use supersonic aircraft, are aware. Sound travelling at 767 m.p.h. takes about five seconds to travel a mile. This is why a man watching a cricket match from a distance *sees* the batsman hit the ball at almost exactly the moment he does it (so extremely swift is the speed of light). However, he *hears* the sound of the bat striking the ball some seconds later. For the same reason one sees the flash of lightning of a distant storm at almost the same instant that the lightning strikes. The clap of thunder comes later, depending on how far away the storm is, because the sound – as is obvious when one thinks of it – can only travel from the point of the flash to one's ear at the speed of sound.

But although the low notes and the high notes, the beat of the drum and the squeaking of the piccolo in a distant orchestra, will all travel towards one at the same rate, the pitch of the note and the speed of the sound can under certain circumstances affect one another. The simplest example is provided by a moving railway locomotive. In a stationary locomotive, held up by signals, let us say, the engine driver hoots at the signalman. This means that the hooter sets up a steady pulsation in the air at, for example, 2,500 pulsebeats a second. The sequence of rapid vibrations in the air, amounting to a series of rapid compressions and expansions, is interpreted by the ear as a certain note. These vibrations travel towards one's head at a steady 767 m.p.h. like a regiment of soldiers marching in step until they have all passed. This is what happens when the source of the sound – instrument, voice, bell or locomotive hooter – and the ear of the listener receiving it remains still. On the other hand, if the thing making the sound is travelling towards the listener, the evenly spaced vibrations in the air will become crowded together. That is to say, when the pulses arrive, more of them will strike the ear than the 2,500 per second which set out at the beginning and the note the listener hears will be higher. On the other hand, if the source of the noise is travelling away from the ear that hears it the pulses in the atmosphere will become more widely separated and the note will sound lower. The full effect is most dramatic when an engine, hooting steadily all the time, approaches at speed a listener standing on a railway platform. The note of the hooter is higher as the locomotive approaches. As it passes, in a flurry of noise, the note suddenly drops down the scale.

This brings me to the answer to a curious question that I was asked by an observant musician. This gentleman, a Mr Heald, was the conductor of a Salvation Army band. He noted that when his players hurried in with their brass instruments straight from work on a cold night, the music they made when they started to play was always flat – the notes they sounded were lower in pitch than when they were warmed up. Mr Heald asked, 'Why does a cold horn play flat?'

The answer is interesting and in some respects unexpected. Everybody knows that metal gets longer – that is, it expands – when it is heated. It follows that the length of tubing in a trumpet or a cornet will become *shorter* when the instrument is cold. But this does not explain why it plays *flat*. If anything, one would expect it to play sharp and produce a *higher* note. The whole principle of a trombone, for example, is based on the fact that when the trombonist wants to produce a low note he pushes the slide away from him and thus makes the effective length of piping *longer*. On the other hand, when he blows a high note he pulls the slide towards him and makes the instrument shorter. The effect is to bring the pulsations that he makes with his mouth nearer together so that their frequency when the air carries them to one's ear is greater. So far as the cold brass instruments in the Salvation Army band are concerned, the amount by which the cooling shortens the tubing is so little that its effect on the note would be virtually undetectable, and in any event, as I have said, it would tend to make the note higher rather than lower if it exerted any effect at all.

No. The cooling has its significant influence on the air inside the horn rather than on the horn itself. Sound, as has been described, travels through the air as a series of pulses. The first lot of gas molecules of which the air is composed bump up against the next, which in turn cannon on to the next and so the series of oscillations travels on, something like the ripples moving outward from a stone dropped into a pool of water. The *speed* at which the series of ripples moves is affected by the temperature. At an ordinary temperature of, say 21°C (70°F) the speed of sound in air is about 767 m.p.h. But when the air is cold it becomes more viscous, just as golden syrup does when cooled, so that at freezing point (0°C or 32°F) sounds travel 27 m.p.h. or so slower, at about 740 m.p.h. The effect, therefore, is as if the horn player were sitting in a train travelling *away* from the listener at 27 m.p.h. That is why the notes sound flat.

Experienced wind-instrument players know that it is difficult to play in tune when their instruments are cold even if they do not always know why. But there is a more interesting example of the effect of the speed of sound on its pitch. In this instance it is a practical example of how an *increase* in speed raises the note. The phenomenon is called 'Donald Duck voice' and it unexpectedly turned up as a practical problem of some import-ance among deep-sea divers working on the North Sea oil-rigs. These men obviously need to talk to each other when they are working together. They do this with two-way radios inside their helmets. But when they first tried, they could not understand what the other man was saying. Instead of manly baritones and tenors booming out inside their diving suits, their voices were squeaky falsettos like nothing so much as Donald Duck in his most inarticulate mood.

The reason was this. Normal air is composed of a mixture of two gases, one part oxygen to four parts nitrogen. In a glass of whisky which is one part whisky to four parts water, it is the whisky that has the effect, the water serving merely as a neutral liquid in which the whisky is carried about. In air, the oxygen is the operational fraction one needs to keep going while the nitrogen is the vehicle in which its travels. One particular characteristic of nitrogen is that, as with carbon dioxide gas in beer, it dissolves in blood under pressure. This means that, just as the bubbles come frothing out of beer when the cap is taken off the bottle and the pressure relaxed, so also, if a man has been taking air into his lungs under high pressure, bubbles of nitrogen gas will form in his blood and froth out when the pressure is abruptly released.

Divers working in deep water are under pressure. This means that the air pumped down from above for them to breathe must be under pressure too. And because this is so, part of the nitrogen gas from the air dissolves in their blood. The men know from bitter experience that if at the end of their shift they come quickly to the surface, which has the effect of quickly relaxing the pressure under which they have been working, bubbles of nitrogen will form in their blood. Such bubbles give them intense pain and may, indeed, kill them. The pain and danger of nitrogen bubbles are called by divers 'the bends'. Until recently, the only known way of dealing with the situation was to relax the pressure very slowly. Although this was effective it possessed its own built-in disadvantages. For example, a man who had exhausted himself working a long, hard, cold stint was compelled to spend some more hours slowly and stage by stage coming to the surface to allow his blood to decompress.

The problem was most severe for the divers at the North Sea oil-field who were working at the limits of depth at which ordinary diving can be done. It was, however, solved in an ingenious way. As I have already pointed out, the nitrogen gas making up four-fifths of the air is quite inert. It was decided, therefore, to make up a 20 per cent mixture of oxygen but using another inert gas, helium, instead of nitrogen. Helium is not only inert but it is also insoluble. And because it cannot become dissolved in the divers' blood whatever pressure they are subjected to, it cannot give them the bends when they come to the surface and the pressure is released.

The use of a helium-oxygen mixture in place of air solved all the deep-sea divers' problems except one. Helium is a lighter gas than nitrogen. It is, in fact, used to fill airships. And because of its lightness and liveliness the ripples of sound travel through it more rapidly than they do through air. That is to say, the speed of sound is greater in helium than it is in air. The consequences are obvious, once one begins to think about it. Just as a cold cornet, tuba or horn plays flat because the speed of sound in cold air is slower than it is in warm air, so the opposite effect occurs when the speed of sound is increased, as it is in an atmosphere of helium: the note is raised in pitch. Had the divers been playing musical instruments at the bottom of the North Sea, their instruments would have played sharp. As it was, the pitch of their voices became raised and an ordinary man speaking, as he thought, in his ordinary voice, sounded like Donald Duck.

The effect of helium can easily be demonstrated without going to the bottom of the sea in a diving suit. If a man takes a deep breath of helium from the nozzle of a cylinder of the compressed gas and then speaks, he will startle himself – and everyone else who is near by – by discovering his voice coming out as a high-pitched falsetto. The effect is even more remarkable for a woman. Her voice will sound like a poor imitation of Shirley Temple as a little girl on the Good Ship Lollipop. Because the sound is travelling faster through the helium than it would through air, the number of pulses per second emerging from the voice-box is greater and the pitch of the note higher. The effect is the same as that of a tape recording playing too fast. And this is exactly the way the Donald Duck effect is produced.

The odd phenomenon of the high-pitched voices of the North Sea divers presented the oil-rig people with an interesting problem. Since, when left to themselves, the divers could not talk to each other with much hope of being understood, something had to be done. The solution was achieved by modifying the microphones and related telecommunication system they used and incorporating an electric device to lower the pitch of their voices in each other's earphones to the normal value.

Of course, it is a good deal easier to play tricks like this with a microphone, which could be described as a mechanical ear, than it is with a real ear. A microphone can be designed to pick up the pulsations in the air which cause sound at any frequency the engineer desires. And those pulsations can be turned into electrical pulses of the same frequency, made to activate a loudspeaker at the same frequency, and hence reproduce the same sound, or else the frequency can be speeded up or slowed down provided the appropriate circuitry is used. But things are different with a real ear. Although an ear is quite a complicated piece of equipment, basically it serves the purpose of converting the pulsations of sound in the air into nerve impulses of the same frequency carrying messages to the brain describing with considerable subtlety what the pitch and character of the noise was that the ear received. Young, fit people with their ears in good working order can hear very low-pitched sounds caused by pulses passing through the air at fifteen beats per second. Anything slower than this is not recorded by human ears as sound at all. As the beats get quicker, the very low notes of sound gradually rise until, when the frequency reaches 20,000 beats a second, the pitch is so high that again one can only just distinguish it as sound. There are a very few young people who can just hear notes so high that they are caused by pulses vibrating at 23,000 times a second, but this is rare. As people grow older, not only do they tend to have difficulty picking up sound at all, but the range of vibrations they can distinguish diminishes and few people over sixty can recognize as sound high notes with a frequency of more than 8,000 vibrations a second. In fact, they grow to depend more and more on the middle frequencies between about 1,000 and 4,000 pulses a second.

Young people with good hearing should remember this when they feel irritated at having to repeat a remark to their old dad. And perhaps they should also remember that although alligators can only hear notes up to 4,000 vibrations a second (just like old men), cats and dogs can hear notes half an octave higher than the highest note people can hear, while champanzees can hear higher notes still, right up to 30,000 vibrations a second, totally inaudible to us and even to some of our cats and dogs as well.

Then there is the question of the bats and how they manage to fly about in a pitch-dark room without running into wires strung across the room from one side to the other. If ever we begin to feel conceited about our own ability to hear – not merely the ability of young people to hear better than old people but of all of us being able to discriminate between different voices and different pieces of music and some of us even being able to detect whether the music is being played or sung in tune – if ever we begin to feel self-satisfied about possessing such abilities, all we need do to bring us back to a proper state of humility is to remember the great population of bats.

There are very many bats distributed all over the world. And our hearing compared to theirs is little better than a blind cripple with a white walking stick compared to an athletic man with two good eyes. There are more than a thousand different species of bats – little tiny ones the length of whose head and body together is hardly one-and-a-half inches, like flying mice, and great flying foxes with bodies one foot long and a wingspan of five feet. Others again feed on frogs, mice and other bats while a major group called the Mega-chiroptera, eat fruit like monkeys.

Most of the time we men and women, who live much of our lives by day, never think of this other world of the bats, who live their lives at night. Nor do we often remember that as we go about our business, whether by night or day, we only hear a small part of the sounds that fill the air around us. Faster and squeakier than we can hear, the bats are sending out their signals and, with their delicate sensitive ears, are receiving sound signals back. A blind man, tap-tap-tapping with his stick as he walks along the road can quickly learn to know when he is opposite the wall of the next-door house, and when that wall merges into the hedge marking the boundary of the house farther along, and when this leads on to wooden palings outside another house farther on still. He does this by recognizing the different echoes of his footsteps and the tapping of his stick. Sensitive blind people can even recognize the presence of other people as they pass by. They detect the slight deadening in the reflected sound they hear. Bats, however, have refined this sense of hearing and can assess reflected sound to a degree far beyond anything of which human beings are capable.

All the time a bat flies, it gives out several kinds of sound. Most important is perhaps a continuous high-pitched squeaking, so high-pitched, indeed, as only to be audible to other bats. This sound is reflected by all the objects in its environment and the bat, with its sensitive ears as recorders, builds up for itself a sort of 'radar scan' by which it knows where are solid objects into which it might fly in the dark. Its ears are so sensitive and

'. . . while chimpanzees can hear higher notes still . . .'

its reactions so fast that the bat can 'hear' a wire stretched across a darkened room and take avoiding action so as not to fly into it. The bat uses its two ears just as we use our two eyes. In the course of an experiment in which one of a bat's ears was plugged up, it was found that the animal could no longer fly safely in the dark. The sonic range which we do hear can be taken to be what the radio people might call 'the medium-waveband'. Below this, where the beats through the air come slower than we are able to hear, there are 'infrasonic waves': earthquake shocks may be as slow as only one beat or throb each 100 seconds. But vibrating machinery or a motor car or railway train may generate infrasonic waves which we do not hear but which affect us none the less. The vibrations of a ship, for example, may cause giddiness quite apart from the sea-sickness caused by the motion of the water. Travel sickness may be due to infrasonic vibrations which clash with the natural vibrations of the traveller's stomach. It is important for all these reasons that engineers designing hand-held machine tools, motor cars, aeroplanes and space capsules ensure that they do not generate unheard sound in the infrasonic range likely to affect the people who are going to use or travel in them.

And above the comparatively restricted band of vibrations that we hear as sound are the more rapid pulsations in the ultrasonic range. Just as a tuning fork can be made to set the air vibrating at any given rate to produce any audible note that may be required, an electrical device called an *ultrasonic transducer* can be made to produce any desired ultrasonic note. Such notes have been used in the same way that bats use them, to give an echo off the bottom of the sea to enable the depth of water under a ship to be measured. This is called 'sonar'. The same system is used to produce an ultrasonic echo through a piece of metal and thus measure its thickness.

Not only are ultrasonic notes very much higher than anything that can be heard by the human ear but it is also possible to give them much more energy. By focusing a beam of high-intensity ultrasonic waves it is possible actually to 'drill' a square hole through a sheet of glass or steel. Ultrasonic waves generated in a liquid can shake it into bubbles. They have been used not only to shake off dirt from the inside of a tank but actually to get chemical reactions to take place by banging two different compounds together, as it were. The same disruptive effect of this powerful inaudible ultrasound has also been used to sterilize a liquid by breaking to pieces such bacteria as may be present in it.

Just as it is true that up on the moon there is not a sound to be heard, it is equally true that down on earth, although we hear many things in the medium-wave sonic band to which alone our ears are sensitive, we know nothing of the low-pitched infrasonic band or the high-pitched ultrasonic range unless we make use of the new tools that scientific technology has provided.

13 Your illusions are showing

When something that has been puzzling us is explained, we talk about *seeing* the point. And once we have seen the point – what Americans call the 'ah-ha!' reaction – the whole thing appears to be so simple that we wonder why we ever had any difficulty before. 'It's obvious,' we say. That is, when someone has shown us. By the time we have reached adult estate we have been shown so many things that we forget how much of what seems natural and obvious had to be taught us in the first place.

There we are, knowing what is going on around us, rational, sensible people earning our living in the full understanding of how things work. Now and then, however, a particularly perceptive individual comes along possessing what might be called 'double vision'. Such a person sees what we see and interprets what he sees the same way as everyone else; at the same time he looks more critically at what everyone else sees and then – without anyone to tell him – points out that what we all believe we know may, in fact, be interpreted quite differently. First-rate scientists are people of this kind.

Let us start with something simple. When you look back along a straight road, you know perfectly well that half a mile back the road is exactly the same width as it is at your feet. A surveyor making a plan of the road would draw it just like that; the same width at point A (where you happen to be standing at the time you look back) and at point B, half a mile away. And he would be right to do so. Not all that many centuries ago people used to draw pictures in this way too. Children, if left to themselves, still do. But when you look back along a straight road, the two sides *appear* to get closer and closer together until, in the far distance, they appear to meet.

Although we know that roads don't really get narrower or men get smaller the farther they are away, this misleading appearance can be useful. For example, it can give one a pretty good idea how far away things are. Having learned to judge distances by the size things appear to be, we are happy and forget that what we are looking at is appearance, not reality. We feel we can cope. That is, until some particularly perceptive person – a scientist, perhaps, with double vision – points out how fragile our ideas of reality may be.

Right up to the fifteenth century, painters painted a square block – whether it was a building or a block of wood – just as it is (or as well as they could do it) – showing its equal sides equal, like this:

Not until an Italian painter, Paolo Uccello, had discovered *perspective*, that is, the way to paint things as they seem to be, rather than as they are, did people begin to draw cubical blocks of wood like this:

Because we have been trained to do so, people like us, when they look at a drawing like that, 'know' that it is a cubical block with equal sides all round. They know this, not *in spite* of the fact that the sides supposed to be farther away are actually drawn shorter, but *because* they are. 'Very well,' one might say, 'so they're shorter. But if the thing *seems* to be a cube, does it matter so long as everybody knows that it is meant to be?' But do they?

Suppose we now set out to draw a transparent block of glass. We should do it like this:

But would *everyone* then know what they were looking at? The answer is no: some of them would believe they were looking at a cube with one of its flat vertical sides facing the right side of the page, while others would think they were seeing the bottom part of a truncated pyramid standing flat on the paper, with its top pointing up from the page. Perhaps we have learned our lessons about perspective too well. As schoolchildren we have drawn those railway lines dwindling away into the distance too often until we have come to believe that the appearance is the reality. Learned mathematical equations have been worked out so that, if you want to, you can calculate the angle of two straight lines getting nearer and nearer, or the still more complicated lines representing curves and irregular shapes as they appear in perspective. The fact that pictures can be painted so well or, for that matter, television images can be thrown on a screen that are so realistic that we would almost swear – if we did not know otherwise – they were genuine horses galloping across the West, does not alter the truth that they are flat images on canvas or on the nearly flat face of a television tube.

There is a simple demonstration – so simple, indeed, that one can hardly believe oneself fool enough to have been deceived by it until believe it one must – which shows how completely we have learned, from what we were told as children and have picked up since then, to believe in appearances that conceal reality. The demonstration can be done like this. First make a trapezoid out of cardboard or plywood and, to make the experiment more elegant, paint it white on both sides to represent what might be taken to be a perspective representation of a window-frame. If this flat piece of board, from which the 'window' openings have been cut out, is mounted on a rod, the rod fixed into a stand (taken, perhaps, from a desk lamp) and the stand stood on the turntable of a record player, the trapezoid can be made steadily and slowly to revolve. But an observer, knowing perfectly well that it is going round and round, cannot *see* it doing so. And because he cannot see it he

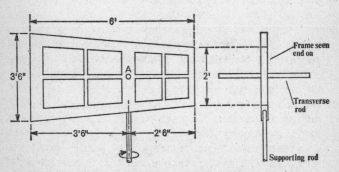

When this structure revolves the narrower end always appears to be farther away than the wider end. The eye is so completely deceived that the whole thing appears to be flapping to and fro when one knows it is actually going round and round. A rod pushed through the hole A (viewed from the narrow end in the right-hand drawing) appears to bend and pass through the revolving frame.

cannot *believe* it even though he knows it is so. The trapezoid does not appear to be revolving; it appears instead to be flapping gently first to the left, then to the right. The eye of the beholder, assuming that it is an 'educated' eye, all the time assumes that the shorter end of the trapezoid is always farther away than the taller end. So firmly does one's mind hold to this belief that if a light rod is pushed through the hole marked in the 'centre frame' of the window so that it sticks out equally far on each side, the observer watching the whole thing as it revolves would swear that the rod was bending and passing through solid parts of the structure.

Everyone sets up a scientific system for himself or herself. That is to say, we all form in our minds a reasonable explanation of the way things work out in the world and what kind of a world it is. The science that pupils learn at school merely goes a little deeper into the various aspects of things which those who studied them before worked out as best they could. These systems are as satisfactory as they can be made but, as the examples I have just given show, though good enough for most of the everyday affairs of life, they may contain weaknesses. It is in these weaknesses and the possible explanation of how current belief falls short of the truth that the *real* scientist with his 'double vision' can sometimes spot.

Ordinary people believe they see the trapezoid 'window frame' flapping backwards and forwards. It would take a near genius, seeing it for the first time, to reason out that it was really going round and round. Not everyone can make out the truth of what seems to them to be nonsense, for example, what is the matter with the diagrams below?

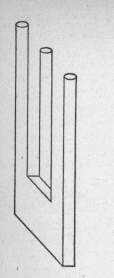

Because we have become so completely 'brainwashed' to believe that a flat two-dimensional drawing on a piece of paper is a solid three-dimensional object, we become troubled when we don't see the drawing we expect.

C. M. Turnbull[1] lived for more than a year with the pygmies in Central Africa. These people are highly intelligent and keep to a quite complicated social system, hunting and camping, getting married and trading with the Negroes who live on the borders of their territory. They spend the whole of their lives in the dense rain forest which is their home. Towards the end of his stay Turnbull took one of the leading pygmy hunters with him on an expedition. Eventually they came to the edge of the forest where the trees end abruptly and open farming country begins. There, across the valley on a distant slope, a herd of cattle was grazing. The pygmy hunter was amazed. First, he took them for insects. Then he exclaimed that he had never before seen such tiny little pygmy cattle. Having lived all his life in the dense rain forest of the Congo, where vision is

1 Turnbull, C. M., *The Forest People*, Jonathan Cape, London, 1961.

restricted to 100 yards or less, he had never before seen cattle *in perspective* and did not believe them to be true.

How sure are we that we know that what we see is true? Clever women know perfectly well how to deceive. A dumpy girl can make herself look quite tall and slender if she wears a well-designed dress that suits her, while a long skinny woman will have learned by experience, if she has her wits about her, how to give the public the impression that she is not really all that much out of the ordinary.

There is little doubt that line A (which should form the basis of the dumpy girl's dress) appears to be longer than line B (highly recommended for very tall girls). In truth, both are exactly the same length.

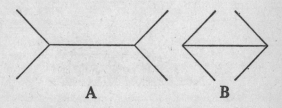

A   B

This sort of thing is much more than a little bit of fun. It lies at the heart of science. What is the truth? How can a man or woman know what they are looking at? For years and years men watched the stars marching nightly across the sky. They learned to map their motions, to steer their ships and judge the seasons by the places the stars moved to in their orderly progressions. But among these steadily moving points of light up in the sky were a number of wandering stars – the planets – corkscrewing here and there, sometimes seemingly turning in their tracks to twist round before moving on. All these men 'knew' that the earth was fixed, steady and immovable with the sun, the planets and the stars moving round the solid, stationary earth. And so they tracked their motions. Suddenly, bingo! Along came Galileo and Kepler – and now us – who 'see' that it is not the earth that is still but rather the sun. And now that we have seen and understood the planets (which, after all, go on moving exactly as they did before) suddenly cease to follow a tangled path but can be clearly perceived to be – like us – circling the sun. Remarkable men – great scientists – taught all of us to see like them, with 'double vision'. Was that, they make us ask ourselves, a glass cube or the stump of a pyramid? Is that, they ask of the picture below, my gnarled old mother-in-law looking straight ahead of her or my beautiful young wife gazing to her right?

Mother-in-law is drawn with a hook nose and a protruding chin; lovely wife has a rounded cheek half turned away.

Today the movies have become so commonplace that people forget that there was a time when they did not exist. Yet I can remember as a little boy before World War I being smuggled into the side-door of one of the earliest cinemas that came into existence to marvel at the flickering image of an express train apparently rushing towards me – accompanied by appropriate train music thumped out on a piano. We were amazed and delighted. I had seen a magic lantern before – what magic it seemed – but never a picture of something that actually moved. Now so familiar are we with what was a short time ago a marvel, that we have almost forgotten that the train is not really moving at all and that all we are looking at is a series of still pictures, each one slightly different from the one before, thrown on to the screen one after the other at such a speed that the images we perceive on the retinas of our eyes and record in our brains become merged together in our consciousness.

Sometimes the realism of this technological production of moving pictures is damaged; for example, a moment occurs when the wheels of the stage-coach that comes galloping into Dodge City appear to be going backwards. It is perhaps, more accurate to say that this happens when we are non-scientists (as everybody is from time to time when relaxing after the day's work is done). I emphasize this because as soon as one begins to think seriously about the wheels seemingly to be in reverse, then the sort of thinking that I have called 'double vision' begins. Instead of watching a wagon clattering up to the Lucky Strike Saloon, we are really seeing twenty-four slightly different pictures flashing before our eyes each second. In each of these pictures one of the spokes of a wheel revolving clockwise, let us say, is farther round than it was in the picture before. There is, however, one particular speed of the coach wheel, when a photograph in which a spoke is exactly up and down will be followed by one in which, although the first spoke will have moved forward, the next spoke has moved up to a position just a little bit short of where the first one was before. And in the next picture again, the same thing will have happened. To the eye watching the series of twenty-four pictures following one another, the *impression* will be of a slow backward movement of the wheel. In fact, the eye will not be seeing spoke A gradually moving back and back. What is really happening is that the eye picks up spoke A and then spoke B, which in the next picture is just behind where spoke A was before, and then spoke C, and so on. If the speed of the wheel changes, for example, if the coach slows up as it comes to a stop, suddenly the impression that it is going backwards will be lost and once again we see it as going forwards.

The ability to stand back and *think* about what he believes he sees is very important for a scientist. For example, when a microbiologist looks into a microscope he sees a picture unlike anything you or I are accustomed to. Put a drop of fermenting grape-juice on a glass slide and look at it through a microscope. You see clusters of spherical or balloon-shaped objects, some of which have little spheres or balloons growing out of them. A century or more ago people saw these things and could not quite make up their minds what they were looking at. The French chemist, Louis Pasteur, looked and thought about what he saw and reached the conclusion that these globular bodies were in fact little simple living plants and that it was they that were consuming the sugar in the grape-juice leaving alcohol in their 'excrement'. It was quite an achievement to 'see' like this. The plants were yeast cells. They have neither roots nor flowers so that they are quite unlike the plants one comes across in 'real' life. They do not possess green leaves so that the food they need is more complex than that required by ordinary plants. At the same time, they can take up nitrogen and phosphorus as plants do in the fields. They are perhaps most like mushrooms although less complicated and obviously not so big. It took quite remarkable 'vision', therefore, for Louis Pasteur, seeing what he saw through his microscope, to say they were plants.

Although Pasteur's insight was far more profound than anything we do when we look at the pattern below, there is something similar about the instant when we suddenly 'see' what we have been looking at.[1] It's a Crusader, the top of the picture

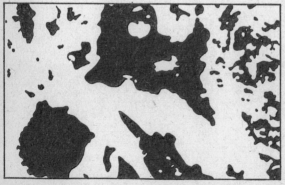

The way a pattern of blobs may fall into shape as a clearly perceived picture of a bearded man was illustrated by P. B. Porter in the *American Journal of Psychology*, vol. 67, p. 550, in 1954. Not everyone finds the man easy to see at first, yet newspaper pictures made up of dots give no trouble.

1 With acknowledgement to Porter, P. B., *Amer. J. Psychol.*, 67, 1954, 550.

cutting across his forehead, and a sheepskin cloak across his shoulders.

This was Part One of Pasteur's 'double vision'. Part Two came when he was asked what went wrong when grape-juice, instead of being turned into agreeable and drinkable wine, went sour. Again he looked through his microscope at a drop of the sour wine. All the other scientists of his time looked too. Naturally, they all saw the same things. These were the usual clusters of yeast cells in the normal background of scraps and specks that one inevitably sees when looking at anything on a microscopic slide. Everyone who looked saw all this. Louis Pasteur alone, however, saw things differently. There were the globular objects, some with buds on them, some without. These were the yeast cells – as all good scientists in the 1870s were then agreed. Very much smaller than these, less rounded or, indeed, not rounded at all, were the general debris, the dust and fluff made visible by the strong magnification of the microscope. No one bothered about this stuff. The pieces in the background of the yeasts seemed too small to be anything else than rubbish. Pasteur, however, saw differently from everyone else. He was the first to suggest that these much smaller particles might also be living cells. In spite of their smallness, only just perceptible through the microscope and thus different from the yeasts, he perceived that they possessed a regular shape. He called them 'rods'. Later their shape was picked out better when dye was put on the microscope slide by which they became visibly coloured. The important point, however, was that having had the idea that these things were living too, he could reasonably argue (and he was right) that whereas yeasts after consuming sugar excreted alcohol, these other things (they were bacteria, not yeasts) after making use of sugar excreted acid, just as the bacteria that make milk go sour do. And from this idea came the enormously important one that just as the bacteria that caused beer to become sour could be described as causing 'sickness' in beer, so also might other bacteria cause sickness in people.

A quick eye and, better still, a quick receptive mind can lead those with 'double vision' to a new understanding of what they see. Few of us possess such insight, which explains why geniuses are rare. But we can watch and think as we move through life; after all, we have problems to deal with too. If what everyone has been accepting as right has actually been wrong all the time, it could be a simple, clear-eyed person like you or me who picks out the error. Look at the three statements below for example:

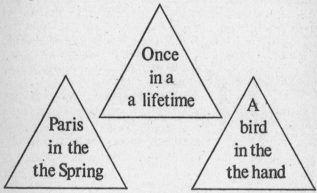

The important results that can follow the flash of insight are well illustrated by Alexander Fleming's discovery of penicillin. To start with, many people would have said everything was against him. He was working in a small unsatisfactory laboratory in Praed Street, Paddington, which itself was not a particularly salubrious district of London. Secondly, he had little help. There was, for example, nobody to wash up the dishes he was using for his experiments. The room was stuffy yet when he opened the window, there was a draught. And the man upstairs was working, under equal hardships, on the study of moulds. Yet out of all this came a first-rate discovery.

Fleming was studying *Staphylococci*, which are the kind of bacteria that give you boils. He introduced a drop of infected liquid containing a few hundred of these bacteria into a circular glass dish containing melted gelatin. The idea was to cover the dish, keep it warm in an incubator and see how each of the bacteria would grow into a 'colony' which would become visible as a white spot on the gelatin. One of these experiments went wrong. What with one thing and the other, Fleming neglected to cover the dish properly, so that, when he went to look at it, a great patch of mould (actually it was the kind of mould being studied by the man upstairs which had floated in through the window) was growing in the dish.

An ordinary man looking at such a dish would have seen nothing other than a spoilt experiment. Had he worked in a well-equipped laboratory he might not even have seen the dish at all: an assistant engaged for the job would have washed it up and set up a new experiment. If this had happened, penicillin might never have been discovered. Fleming, however, did two things. He really looked at the dish and, having done so, he thought about what he saw.

What, in fact, did he see? On one half of the dish he saw the colonies of *Staphylococci*; on the other part he saw the patch of mould. But he saw more than this. He saw, in fact, what was *not* there. And what was missing was any scatter of *Staphylococci* near the mould. Why should there be? The *Staphylococci* were there first before the mould blew in. Why then had they not grown anywhere near the mould? There are several possible answers to this question. One answer could be that the mould was releasing something into the gelatin that hindered bacterial growth, i.e. that it was capable of producing an 'antibiotic'. And this, as was shown later after a good deal of further detailed work, turned out to be the truth. The undistinguished, even messy sight presented by a growth of mould on a dish of infected gelatin where it had no right to be to a man with the imagination to spot the inner meaning of what he was looking at, contained the secret of penicillin, the first and best of the antibiotics by which millions of lives have since been saved.

Fleming's discovery of penicillin is an example of the good that can come out of the ability to look and to see and to understand, unperturbed by what other people think they see. Mark you, there is never any harm in science, if by science we restrict our meaning (as it should be restricted) to the understanding of the way things are in Nature. The trouble arises when people use their knowledge unwisely or – to put the same thing the other way round – when they fail to use the knowledge that is available. Think, for example, about the way in which we judge the size of an object by comparison with its surroundings.

A film set was once constructed for a story in which the actors were supposed to have become as small as mice. The producer simply had part of a staircase built with each step about nine feet high spread with carpet one foot thick. There is a more interesting phenomenon. Two men, both the same ordinary size, can be made to appear like a giant and a pygmy doll together. What misleads the eye of the observer is that a room is constructed with the dimensions all wrong. On the left side the door is about twelve feet high and, at the same time, the width of the room is much greater than it is at the opposite wall. The effect is to make the man appear to be very small. Again, the back wall is not, as it appears, the same height on both sides. As one moves to the right, the ceiling comes lower and lower. The reason why this is not noticeable is because at the same time, the floor is neither square nor level and the right-hand corner, besides being lower is nearer to one's eye. Therefore, when a man comes through the right-hand door, where the ceiling is only about three feet six inches above the floor, he *appears* to be enormously tall.

This whole business may seem like fun and nothing else. But it is much more. We can only know by what we see, together with what we feel, hear, smell and taste. Seeing is, however, the most important. As the saying goes – and the saying is remarkably true – seeing is believing. By seeing, thinking and then reasoning, real insight into what the world is all about has been obtained and, very often, scientific discoveries leading to great achievements have come about.

14 Ask a silly question and ...

Given half a chance, most of us will ask the questions that have been asked since time began. These questions have nothing to do with the balance of payments, the energy crisis or industrial efficiency, all of which are undoubtedly important. Rather people are asking *still* – as they have always done – why is the sky blue? Where does the wind come from? Where does it go? And why can we not see the air as it brushes our cheek? Come to that, there are people who ask – aye, who write letters and stick expensive stamps on them to get an answer – why we can see through a plate-glass window but not through a brick wall.

(All right, Clever Dick, what is the answer? Or have you never considered the matter? After all, both the plate-glass window *and* the brick wall are made from much the same stuff – sand! And what about H. G. Wells who wrote one of the earliest science-fiction stories, *The Invisible Man*, about a man who makes himself like glass and who could streak about the town unseen even though he found it difficult to run down stairs, not being able to see his own feet?)

Some of the simple questions non-scientific people ask their scientific friends are, in fact, quite simple and quickly yield to thinking, particularly when the explanation has been worked out by a clear-thinking scholar of the past. For example, one of the questions I am asked more often than any other is: 'Why does one's face appear upside down when reflected on one side of a silver spoon and the right way up (although not a flattering likeness) when reflected on the other?'

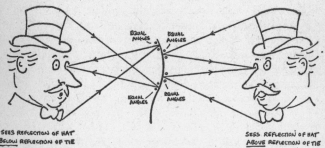

SEES REFLECTION OF HAT BELOW REFLECTION OF TIE

SEES REFLECTION OF HAT ABOVE REFLECTION OF TIE

Although it does seem odd to see one's face the right way up – though paunchy – when reflected from the back of the spoon and the wrong way up when reflected by the inside of the spoon, the answer is easy as soon as one grasps the fact that light bounces off a mirror like a ball bouncing true off a wall. To see the reflection of one's own eye, the light from the eye must be bounced fair and square or, as the scientists say, at right angles, off the mirror. For this to happen *inside* a spoon, the bounce must be off a lower part of the reflecting surface. To see one's chin, the light must bounce from an upper part of the surface if it is to reach one's eye. The whole image, therefore, as the drawing shows, appears upside down. This is not so for light bounced off the reflecting surface on the back of a spoon (or better still, off a flat mirror).

It is nice to know the answer to problems, especially when the answer is neat and tidy and – as they say – intellectually satisfying. And a great deal *is* known about all sorts of things, which is why television sets and tape-recorders, aspirin tablets and 'instamatic' cameras are possible. But it is also nice to know how strange and mysterious the world still is: how much more there is to know and how often a 'silly' question is not silly – in the ordinary sense of the word – at all. 'The silly sheep look up and are not fed,' says the poet, referring to the words of Holy Writ. Here he uses the word 'silly' to mean 'simple'. And a simple question is always worth asking by those who want to know. Sometimes the answer is simple too! We now know why we're upside down when we look at ourselves in a spoon. But when we look up to heaven on a fine day, why is the sky blue?

There are almost too many scientific answers to this question for all of them to be entirely believable. Lord Rayleigh, the great physicist of a century ago, had a lot to say about it. Clearly, the colour of the sky has something to do with the atmosphere. To astronauts in their space-capsules, high above the atmosphere, the sky is jet-black. But if the molecules of air together with the molecules of water mixed with them produce a colour by scattering the light, why is the wavelength of what is scattered that of *blue* light? Is there any reason why it should not be green or yellow occasionally? And why the deep changing shades of blue, pale and haunting in northern latitudes, strong and hard in the tropics?

The blueness of the sky is not entirely due to the molecules of oxygen and nitrogen, the molecules of ozone formed high up, the molecules of water and the minute globules of fog; dust has something to do with it too. The colour is affected by the dimensions of the particles, whether they are dust particles or particles of smaller molecular size, in proportion to the wavelength of light.

Light as it comes from the sun is a mixture of all the colours of the rainbow. Light is energy and it can be thought of either as tiny pellets of energy, photons as they are called, or as vibrating waves. Whichever way it is looked at, light rushes towards us vibrating up and down and the length of each wave, from one crest to the next, is shorter when the light is mauve or blue than it is when the colour of the light is red or orange. In a vacuum, where there is no atmosphere, the light shines hard and true – and the sky, wherever there is no light, is jet-black.

But in the atmosphere the light strikes the floating molecules of gas and the tiny particles of dust, smoke, pollen or bacteria and is reflected and diffused; just as light shines *through* clear glass, it is diffused if the glass is roughened by being frosted. With the atmosphere, where the size of the particles against which the particles of light bump is very small, the light which comes oscillating with a short wavelength is more disturbed and bounced about than the light which has a longer wavelength. So the long waves, which bring the heat of the sun and the infra-red rays that make one brown, pass right through the atmosphere and hit the ground (or us, if we are there). The colours red and orange and yellow mostly pass through too. But as the coloured light that travels with the shorter wavelengths comes along it is much more diffused; the atmosphere behaves like frosted glass for them. It follows, therefore, that as we look up to the sky the diffused light we see is mainly blue.

In 1883 there were times, all over the world and most particularly early in the mornings and late in the afternoons, when the sky was not blue. Instead it glowed with brilliant unearthly colours – reds, yellows, even greens and sombre purple. These changes in colour were due to the presence of particles of dust from one of the most overwhelming phenomena ever recorded on earth. It was on the morning of 27 August 1883 that three of the most violent paroxysms of a volcano, which at that time was 2,623 feet high, on the small island of Krakatoa, then situated in the Sunda strait of what is now Indonesia, took place. The whole northern portion of the mountain and the island on which it stood were torn away. Where a few moments before had been the slopes of a peak rising into the clouds was a vertical cliff which laid bare the inner structure, the very bowels of the volcano. Instead of an island six miles long and three miles across rising at its highest point to the mountain peaks, there was a submarine cavity, a great hole in the sea bottom 1,000 feet deep into the glowing entrails of the earth into which poured the cold waters of the ocean, to explode with shattering force into the sky.

For two days great convulsions took place. One hundred miles away the sky was darkened at noon. A column of stones, dust and ashes was projected into the air. Pumice-stone bubbled out of the crater like froth from a bottle of beer and floated on the sea in 'mattresses' six feet thick. The dreadful sound of the eruption was heard at Roderiguez nearly 3,000 miles away, in Ceylon, 2,000 miles away and in Australia, 1,500 to 2,200 miles away. Worse still were the great tidal waves which drowned 360,000 people and even circled round the world to the English Channel. For a time in 1883, when Krakatoa had thrown fine particles of dust into the high layers of the atmosphere where they were carried all round the globe, the sky was no longer blue.

An even better question is 'Why is the sky black at night?' Ask a scientist and he will tell you that up there in the sky light can travel on at the tremendous speed it goes for years and years – for ever (why not?). Ask him again and he will say that wherever you look on a clear night, in whatever direction you direct your eyes there is bound to be a star, whether it belongs to the millions in the galaxy to which we belong or to some of the millions of other galaxies all about us. Why then, you may well ask, instead of the night sky being black, is it not a solid arch of shining silver?

It's a simple question based, I hope you will say, on common sense. And its answer is simple too, in its way, a 'Super-Krakatoa' one. The sky is black, because at the Beginning[1] (the scientists say) there was an explosion and everything is flying apart still – and the outside stars were shot off so fast that they were out of sight before we saw them (or ever shall).

Another good question is the one I started with: 'Why can one see through a plate-glass window but not through a brick wall? After all both glass and bricks are made of sand, aren't they?'

All the solid material we see around us is, in the main, not there. What I mean to say is that the smallest bits of stuff, the unit portions, so-called *atoms*, are themselves a nucleus – a tiny sun, as it were – encircled by a 'solar system' of whirling 'planets', the electrons. This implies that, like the great solar system in which we live, the little solar system of an atom is

1 Even the best of scientists have the grace to be a bit hazy when this was. Most of them think, however, that things started somewhere between 8,000 and 60,000 million years ago.

mostly space. More than this, however, a grain of sand, a lump of rock, a diamond – or a piece of glass – possesses a crystal structure. That is to say, like a honeycomb, it is made up of a regularly designed system of scaffolding. Different materials – iron, ice, cement, fingernail parings – are obviously made up of different stuff fitted together according to different patterns. Sand, quartz, flint and the like are all made in the main of the element, silicon. When it is heated red-hot and then cooled, the silicon atoms fit together to make glass. Although glass is hard, the pattern of its atoms (if one could become small enough to look at them) is in fact a comparatively openwork pattern. And if the surface of the glass is polished smooth, photons of light, no matter at what wavelength they are vibrating, will pass unobstructed right through. Some glass has, here and there, atoms of lead caught up in the network of silicon. This fits in without obstructing the photons of light as they flash through. But if some iron or a little copper or cobalt is added to the silicon, photons vibrating at particular wavelengths can pass through but the others cannot and the glass, though still transparent, is coloured.

But what about the brick wall? It's mainly silicon too. There are two answers why light will not go through it. The first is that bricks are rough. If one made glass bricks as rough as building bricks one would not be able to see through them either, because the light would be scattered by reflection off all the little roughnesses. The second answer is that the crystal structure of the bricks is also wrong. Grind up the bricks and melt them in a furnace and one might still be able to make something like glass out of them.

It is not a yes-or-no business whether one can see through something. The photons of light, whizzing along at the tremendous speed they travel and oscillating up and down, in short waves if they register as violet or blue and in longer waves if they register as orange or red on the retinas of our eyes, can penetrate anything at all. The point is that sometimes, if the scaffolding structure and the massiveness of the atoms is too much for them, they do not penetrate very far and most of them either give up their energy there and then or bounce back. The first step to take if you *do* want to see through something is to make sure that the surface is smooth. With glass, as I have already said, one must make sure the surface is polished, not roughened as it is in frosted glass. If one wants to see through a piece of leather, a piece of parchment, for instance, one can do so moderately well by rubbing oil on it. The same also goes for paper. The oil makes the surface smooth so that the light does not get scattered.

One can see through some stones, too, provided the surface roughness is polished away. It is, for example, quite possible to make a 'glass' for a watch by splitting off a slice of quartz and polishing it. The light can find its way between the atoms much as it does through glass.

The plastic used to make the windows of helicopters and the like is interesting stuff. Once again one can see through it because most of the light's photons wriggle through the network of atoms of which it is made without much trouble no matter what their wavelength is. But if the plastic is strained or put under pressure its crystal structure can alter and it may become opaque. The photons of light become tangled in the meshes and one can no longer see through it. This change due to mechanical stress can be used in rather an ingenious way. An architect or engineer who wants to find out where the main stress is likely to fall on the supports of a large building of new design – a cathedral, for example, or perhaps a multi-storey block of offices or a deep-sea oil-rig – will build a model out of transparent plastic and then lean on it in order to imitate the maximum stress the real building would ever have to resist. Where the transparent plastic has become non-transparent – that is, where the molecular structure has changed – are the places where the stress is greatest.

But what about flesh and blood? Was H. G. Wells right and is an invisible man possible? Although the answer to the last question must be no, it is a more subtle matter than one might think. After all, we have already noted that one can see through oiled parchment, which is the skin of a calf. Human skin is also transparent up to a point and some people's skins are more transparent than others: one can easily see veins through the skin of pale, delicate people. And what about eyes?

Everyone reading this book is looking through two crystal clear lenses like shining, polished, clear-glass marbles, and then through two clear protective 'lamp glasses'. These are the corneas – the front of each eyeball. Or, to put this statement the other way round, light has no difficulty in passing virtually unimpeded through a first protein structure, the cornea, and then through a second, the lens, to get to the nerve endings on the retina (the 'camera film', as it were) where what is seen is telegraphed to one's brain. Flesh is, in the main, made of protein just as one's eye lenses and corneas are. It so happens, however, that the way the cornea and lens proteins are put together allows light to pass whereas the way muscle protein is assembled does not.

How subtle the whole business is becomes apparent when one considers the eye disease, cataract. Cataract often turns up – no one knows exactly why – in older people, or in people with diabetes, or after taking certain types of drugs. And what happens is that the lens in the eye of the sufferer is no longer clear and bright like a shining glass ball. Instead, here and there in it are patches of opacity, as if some of the glass inside the marble had become frosted (or as in the plastic model of an oil-rig that has been subjected to sufficient stress to rearrange part of its molecular skeleton). But as I said, the reason has yet to be discovered.

Although certain tyes of protein structure – in the human eye, in certain kinds of water-fleas, and to some degree in skin as well – are transparent, blood is more difficult to see through, as any boxer with a cut over his eye will tell you. This is because some of its molecules have a complicated structure built round an atom of iron. Light of shorter wavelengths cannot get through this; longer wavelengths do, and that is why we see red when looking through blood.

Of course, one can see through people if instead of using light one uses X-rays, whose wavelength is about 10,000 times shorter. But X-rays, like light, can bump into certain molecules, become absorbed or bounce off. While one can 'see' through a person with X-rays, one cannot see through a lead sheet.

There is a way, in theory, by which we could 'see' through anything. Winging in from outer space is a kind of radiation made up of particles called *nutrinos*. These, it seems, can drill their way straight through the earth itself and come out on the other side. So far they cannot be used for 'seeing' at all: scientists only know they are there by tracking their path in a bubble-chamber. But if they ever could be used, everything would be transparent. Meanwhile we have to put up with the fact that we do not see through brick walls.

There's another common question: 'Where does the light go when a candle is blown out?' That's easy. The candle flame, when carbon from the wax is combining with oxygen from the air to produce carbon dioxide gas, is hot. It is hot because the carbon is eager to snap its atoms on to the oxygen. They rush, as it were, into each other's arms and as they combine heat results. Heat is energy and such is the intensity of the heat that some of it flies off in all directions as light. The candle flame is bright because photons of light fly straight into our eyes. The room is light because other photons bang about, bouncing from one object to another so that a few bounced off walls or the ceiling, the furniture or the cat, come too into our eyes and we see the whole room by reflected light. Blow the candle out, the fountain of photons stops and all goes dark.

Or again, 'What is thunder?' somebody asked me. 'And why does it sound so loud?' An easy answer is to say that thunder is caused by lightning. Well, so it is, but this is not a proper explanation. Scientists and non-scientists alike have been asking for ages why it is that lightning, which is the discharge of electricity, should make a noise at all. A number of answers have been put forward by different scientists over a couple of hundred years or more. One suggestion was that as the flash occurs a vacuum is formed and the air rushing in to fill it makes the thunder. Closer study seemed to show this was not so. Another idea was that the flash turned part of the water in the sky into steam and this produced the thunder clap. Further evidence showed this too to be wrong. A third notion was that the water, H_2O, was not turned into steam but into hydrogen and oxygen and that the hydrogen exploding made the noise. Again, a second look showed that this was not right. Modern study seems to show that as the lightning flashes and a big discharge of static electricity takes place, the energy released is so great that a temperature up to 30,000°C (54,000°F) is reached. This causes the air to expand so much and at such a rate that a noise of thunder is produced.

But thunder is not – or at least only on rare occasions – a single crash. As a general rule it growls and rumbles, often for a considerable time. This brings us to another comparatively new discovery. This is that the lightning we see is not a discharge of electricity from the sky above *down* to the ground below. On the contrary, it is mainly electricity in the ground flashing *up* into the sky. What happens is that a cloud becomes charged with electricity, which eventually discharges along a forked path (the line of least resistance) down on to the ground. This discharge, however, is hardly visible. What we do see is the upsurge that happens immediately afterwards when the locally overcharged earth flashes upwards along the same forked path back into the cloud. It has been found that the almost invisible escape of electricity down followed by the lightning flash up may take place three or four times within a tenth of a second. Sometimes it may happen forty times. This is one reason why thunder can be a drawn-out rumbling rather than a short sharp crash. The next reason is due to the fact that the banging may have to travel some distance in getting to our ears from the point where it was actually made. It seems that when a zigzag of lightning occurs, at each point where a zig or zag takes place – that is, where the discharging electricity changes direction – a separate clap of thunder is produced. If we take it – as most people do – as a rough-and-ready measure of distances that each count of one between the time of the lightning flash and that of the thunder it causes represents one-fifth of a mile, then *from a single flash* there will be five seconds between the thunder due to a zig occurring just above one's head and that from a zag 5,280 feet (a mile) up in the air. In other words a growl of thunder is due partly to repeated flashes up and down taking a tenth of a second to happen and to noises produced at each corner all the way up the fork of lightning.

There is another interesting thing to notice about thunder: whereas in the main it rumbles and bumps and growls as the lightning flashes and the rain pours down, every now and then there is a terrifying crash that sounds quite different from what has gone before. As it happens, this sort of crash – which can be described as a *clap* of thunder – is something that one is justly terrified of. The reason it sounds different from the rest of the storm is that it is *nearer*. Although low notes, middling notes and high notes all travel at the same speed (the speed of sound), the low notes travel farther than the high notes. Distant thunder reaches one's ear some time after one has seen the flash of lightning with which it was associated, and since the high notes in its original noise weaken and die away sooner than the low notes, only the low grumbling sound reaches the distant ears. When the flash and the accompanying crash happen quite near, all the notes, high and low, reach one's ears. The thunder then is sudden, loud, and at a higher alarming pitch.

While scientific study seems to have explained what thunder is and why it sounds the way it does, there is still some uncertainty about how clouds get charged with electricity and why the electricity oscillates up and down as I have described. There are times, too, when the electric charge at the top of a cloud discharges itself as lightning and thunder with the opposite electric charge at the bottom of the same cloud without ever coming down to earth at all. The method of study at present used is to make a careful recording of the thunder of a storm and work backward from the recording to find out where all the discharges along all the jagged bits of lightning must have been.

Thunder, lightning, blue sky, one's face upside-down in a soup spoon, how to see through a brick wall – all these topics may form the basis of what seem to be silly questions; that is, until one begins to dig into them. Then they can be seen to be deep problems which scientific scholars have thought about for years. And although there may be a great deal of information about all of them, mostly, as with so many other aspects of Nature, knowledge is incomplete and our understanding of what the knowledge means only partial as well. It follows that if you ask two equally wise scientists a question, although they may both know what has been established on the matter under discussion, the way they see and interpret the facts may differ. Facts are sacred to a scientist and reason is sacred too. Just the

'. . . flashing up into the sky.'

same, the scientist is human and part of his or her humanity comes into what he or she understands.

The human element may indeed be as much a part of a scientific question as any other measurement or observation. 'Why is it,' you may ask, 'that when one takes a week's holiday at the seaside, the first three and a half days seem to be much longer than the second three and a half days?' This again is not a silly question even though everyone knows that, if you measure them with a clock, one three and a half-day period is exactly as long as any other three and a half-day period.

In spite of what a clock may say, it is an undoubted fact – and science deals in facts – that to the human consciousness the first part of any event extending over a period of time appears to be longer than the last even though (as measured by a measuring instrument) both are the same length.

Try an experiment. Take a friend for a car drive or a bicycle ride extending over a period of, say, thirty minutes. Let the journey be a twisty one involving diverse changes of direction and scenery unknown to the friend who is with you. At the end of the drive tell the passenger exactly how long the trip has taken. Then ask him to specify the point – the church, the haystack, the traffic lights or wherever he selects – when exactly half the journey time had passed. The average point chosen by any number of people tested in this way is *before* the half-time position. The same result is obtained when people are asked to identify the point at which half the distance had been traversed.

As with experimental journeys of a few miles occupying a time of a few minutes, so it is with life itself. Ask any man in his sixties or seventies and he will tell you that his years as a young man, his early married life when he was in his prime – these occupy the main substance of his life as he looks back. He speaks the truth when, in reply to the address of the chairman giving him a silver watch on his retirement, he says that the forty years of his service with the firm have passed like a flash. Science is the measurement of reality. Is it true, therefore, or not, that *in reality* the first thirty-five years of life are longer than the second? *You* tell *me*!

This question which *I* have just asked *you* is not a silly one. Time is peculiar stuff. We have become so accustomed to our clocks and watches that we forget that just because they *measure* time in a particular way, this does not imply that they provide any real information as to what time actually is. In fact, it's almost an accident, like driving on the left-hand side of the road, that we measure time the way we do. In the early days when the Egyptians were in the forefront of human civilization, their thoughts about time were quite different from ours today. And who is to say that our ideas give a better idea of the truth than theirs did? The Egyptians invented the sun clock. They drew on a horizontal dial the path of the shadow of a vertical stick. Then they marked out the dial of their clock in equal divisions. Each one of these was an 'hour'.

People are not born with any particular ideas about time. A child lives from moment to moment. Night follows day. Up till noon, the shadow of the stick in the centre of the sun clock gets shorter and shorter as the day progresses; after noon it begins to get longer and longer; noon is a mark that half the day has passed. It was, therefore, not unreasonable for the Egyptian horologists to divide the track of the sundial's shadow into exactly equal parts. But the sun passes overhead in an arc, and its shadow passes across the graduation of the dial in different periods of time (using the word in the way we have accustomed ourselves to use it). The Egyptians thus made each of their hours a different length. They also adjusted them to suit the seasons. Their winter hours were one-seventh shorter than the summer ones.

Now here's a curious fact. After some centuries with only sun clocks the Egyptians invented water clocks, in which they told the time by allowing water to run steadily out of a cylinder or some other container. And they went to enormous trouble to invent a series of valves so that the hours measured by the water clocks were of the same irregular length as the irregular hours measured by the sun clocks to which they were accustomed. They even arranged for a slave to change the clock and alter the valve each month so that it told the time like a sun clock all the year round. In describing all this, I am not suggesting that we should give up our equal uniform hours and

181

adopt hours of different length for different times of the day – although I can see a number of advantages in the Egyptian system. I am asking you to recognize two things: firstly, that there are different ways of looking at time; and, secondly, that to people – and perhaps to animals as well – the first part of any event that occupies a period of time appears to be longer than later parts.

This brings us to a question which many scientists not so long ago *would* have thought to be silly, but which some scientists today worry about. It is: 'To what extent does the person asking the question, or the scientist doing the experiment, by the very fact of being there actually affect the answer that best represents the truth or the results of the experiment?' A question I was asked a little time ago, not once but several times and by several people was: 'What does a mirror reflect when nobody is in the room?'

Some people may think that this is no question at all. To them there is no doubt that the universe goes on obeying all the rules of science whether any living human being is there to see things happening or not. It is no good telling such a person that the 'rules of science' would never have been known unless there had been men to discover them. His answer is that the rules exist and go on operating whether they were discovered by human beings or not. 'Yet', the questioner might continue, 'if there were three of us in a row looking at a big mirror on the wall and one goes away, then there are two of us left in the reflection I see. And even if two people move away,' my interrogator continues, 'I look in the mirror and still see one person, namely me, in the reflection. But if we're all three there and *I* move into the next room, what is left of the reflection then? After all, a reflection is something to do with the retina of my eye, so that if I take the eye away, how does the reflection exist?'

Before we begin to answer this question, let us ask another. 'If a tree falls in the desert and there is nobody there to hear the crash, can it be said that the tree has made a noise at all?' Before anyone starts to call this a silly question I should perhaps add that Ronald Knox, a man of considerable intellectual power, went even further. He not only challenged the idea of there being noise when nobody is around to hear it, but he put up the further proposition that it was hard to prove that a tree existed if it lived and died where no one ever saw it. He wrote a limerick on the subject:

There was once a young man who said 'God
Must think it exceedingly odd
If he finds that this tree
Continues to be
When there's no one about in the Quad.'

Years ago I wrote a book called *The Boundaries of Science*. I felt then that the subject was worth thinking about and I still believe it. Science is a delightful topic to study, puzzle over, and ask questions about. It *is* worth knowing which way the water runs down the plughole in the bath, and why the sky is blue. It *is* worth trying to understand what makes perpetual motion impossible. But there are limits to what that kind of inquiry we call science can find out. In other words, science *has* its boundaries. Brought up as we are in the busy world of buses, plastic macs, clocks and radio-sets, we find it natural to believe that the rules of science go on just the same whether we are there or not. If we stop to think for a moment, however, we don't *know* this is so, we only *believe* it. Since everyone on earth, whether he or she is a scientist or not, can only know about thunder and lightning, boomerangs, falling slices of bread and butter, and the smell of rain by what he or she sees, hears, feels, tastes or smells, no one really *knows* whether, when we personally are not there, or when we are fast asleep – or dead – anything is happening at all. As Shakespeare put it,

The cloud-capped towers, the gorgeous palaces,
The solemn temples, the great globe itself,
Yea, all which it inherit, shall dissolve,
And like this insubstantial pageant faded,
Leave not a wrack behind. We are such stuff
As dreams are made on, and our little life
Is rounded with a sleep.

Mark you, common sense, or, as one could say, the *scientific*

183

approach, is to assume that all the scientific laws and rules *do* go on whether we like it or not and – perhaps more important to our peace of mind – whether we personally are there or not. Brought up in an atmosphere like this we have come to assume that there are two quite different worlds. One, the world of science where everything that happens – the stars, the trains, computers and the breeding of baby monkeys – can all be explained by science. And, for the most part, so they can. And a second world, the world of feeling, of believing, of Shakespeare's towers and palaces. In this what we feel, what we are and what we imagine are all-important. When we sleep (or die), this world ceases to exist.

It is strange to find that the advance of science is itself, in a curious way, bringing the scientist – complete with feelings, ambitions and hopes – into the cool, intellectual world of science. There has been a notion that certain things scientists cannot do. One of these is to measure the length of a metal ruler.

They can start off full of confidence and decide the length is twelve inches. Then, to make the measurement more accurate, they scrutinize the end of the ruler through a microscope fitted with crossed hairs and find that the length is a few decimal points different from twelve inches. So, to improve the accuracy still further they decide to use an electron microscope to fix where the ruler finally ends. But as their measuring tools get better and better, they discover that the metal ruler does not in fact have an end at all. When they get down to molecular dimensions, the end of the ruler is a fuzzy cloud of vibrating molecules encircled by their solar systems of revolving electrons.

'Pooh, pooh!' you may say. 'This is only a technical question, and who wants to measure lengths as accurately as this anyway?'

The answer is that some people do, and all of us would if, by succeeding, we could understand better this delightful world and the blue sky overhead. But as we try to do so by using science, science itself drags *us* (with all our dreams) more and more closely into the very fabric of its laws. The very physics and mathematics by which we have got to the moon and now begin to understand the nature of the stars depend on our tracking the positions of individual electrons and measuring the speeds at which they travel. To achieve this, we must shine light on a single electron so that its movement can be observed. But when we do, the photons from the light interfere with the movement of the electron we are watching. So we, the people who want to know, upset what we are studying by the very fact of looking at it.

Isn't it delightful to know that every silly question, if answered with wisdom, can draw us to the ends of human thought? Pull out the bathplug and watch the earth swing round the sun.

Index